Space Law and Policy

Series Editor

Maria A. Pozza, Gravity Lawyers, Christchurch, New Zealand

Space activities across the Asia-Pacific, specifically in Australia and New Zealand, are increasing and diversifying. Over the last decade, countries in this region have seen significant growth within their national space industries and economies which in turn benefit their wider societies due to the downstream application and use of, space-related technologies, products, and services. This phenomena presents not only a great opportunity for these countries, but also represents a situation necessitating considered and fit-for-purpose space law and policy. Robust yet flexible regulatory regimes are needed if countries in the Asia-Pacific, such as Australia and New Zealand, are to benefit from the strategic and commercial advantages associated with a presence in outer space and functioning domestic space sector. These regulatory regimes must respond to, and address, numerous challenges ranging from adherence to international legal obligations to issues of geo-political space security and military concerns through to questions about how best to regulate and support commercial space endeavours. These myriad factors will have an impact upon how countries develop, implement, and refine their space laws and policies. This series offers a unique perspective from this region and intends to act as a source of knowledge and guidance for space actors across government, the private sector and academia in the pursuit and furtherance of their respective space activities and operations. The contributors to this series are all experts in their field who share a common interest in strengthening space governance via the promotion and development of space law and policy.

Hamid Jahankhani · David V. Kilpin ·
Eustathios Sainidis · Maria A. Pozza
Editors

Space Mining

Humanity's Quest for Equity

Springer

Editors
Hamid Jahankhani
Northumbria University
London, UK

David V. Kilpin
Northumbria University
London, UK

Eustathios Sainidis
Northumbria University
London, Northumberland, UK

Maria A. Pozza
Gravity Lawyers
Christchurch, New Zealand

ISSN 2662-9054 ISSN 2662-9062 (electronic)
Space Law and Policy
ISBN 978-981-95-5542-0 ISBN 978-981-95-5543-7 (eBook)
https://doi.org/10.1007/978-981-95-5543-7

This Springer imprint is published by the registered company Springer Nature Singapore Pte Ltd.
The registered company address is: 152 Beach Road, #21-01/04 Gateway East, Singapore 189721, Singapore

Contents

From Terrestrial to Space Mining: Logistics, Regulation, and Transparency Through Artificial Intelligence

David V. Kilpin, Hamid Jahankhani, and Eustathios Sainidis

Abstract Space exploration presents new and extraordinary opportunities for extracting minerals, as celestial bodies contain vast reserves of valuable and rare elements. As humanity stands on the brink of this new frontier, the integration of Artificial Intelligence (AI) and blockchain technology offers innovative solutions to the complex challenges of space mining. This chapter examines how the combination of AI and blockchain can enhance operational efficiency, ensure secure and transparent data management, and foster international collaboration in space ventures. AI can significantly improve the accuracy and speed of mineral detection and extraction through advanced analytics and automation. Meanwhile, blockchain's decentralised ledger provides reliable tools for tracking transactions, ensuring accountability, and supporting equitable distribution of space resources. Entering this uncharted territory demands a robust and integrated technological framework to ensure that space mining is conducted sustainably and ethically.

Keywords Space mining · Governance · Artificial intelligence · Regulation · Terrestrial mining · Minerals · Logistics

1 Introduction

As humanity ventures further beyond Earth, the extraction of resources from outer space has shifted from science fiction to a rapidly advancing scientific and commercial pursuit. Space mining—the recovery of valuable minerals and elements from celestial bodies such as asteroids and the Moon—has become a key area of interest

D. V. Kilpin · H. Jahankhani (✉) · E. Sainidis
Northumbria University, London, UK
e-mail: hamid.jahankhani@northumbria.ac.uk

D. V. Kilpin
e-mail: david.kilpin@northumbria.ac.uk

E. Sainidis
e-mail: eustathios.sainidis@northumbria.ac.uk

H. Jahankhani et al. (eds.), *Space Mining*, Space Law and Policy,
https://doi.org/10.1007/978-981-95-5543-7_1

in the evolving space economy of the twenty-first century. These resources, which include rare earth elements, platinum-group metals, and water ice, offer the potential to both alleviate pressure on Earth's depleting supplies and support sustained human presence and industrial operations in space. However, the immense technical and logistical challenges—ranging from harsh extraterrestrial conditions and high costs to communication lags and minimal human oversight—necessitate cutting-edge technological innovation. In this landscape, artificial intelligence (AI) stands out as a transformative force. AI technologies can enhance all phases of space mining, from autonomous navigation and robotic extraction to system maintenance and decision-making in data-poor conditions. With tools such as machine learning, computer vision, and smart control systems, AI can improve the efficiency, safety, and economic feasibility of off-Earth resource utilisation. This paper investigates the convergence of AI and space mining, exploring how recent advances in AI are reshaping the landscape of extraterrestrial resource extraction and considering the broader consequences for future space missions and resource governance. Several scholars stress the importance of clarifying the current framework to support private sector involvement in space development (Cooper, 2003; Delgado-López, 2015; Genta, 2014; Pop, 2001; Saletta & Orrman-Rossiter, 2018; Shaw, 2013). Cooper (2003) and Pop (2001) specifically argue for establishing private property rights in outer space. Pop (2001) also highlights the necessity of mechanisms to secure such rights.

Meanwhile, other researchers focus on space mining, advocating for regulatory frameworks to govern these activities (Saletta & Orrman-Rossiter, 2018; Shaw, 2013). Beyond legal incentives for space mining, some discussions explore its potential environmental benefits for Earth, as well as the broader issue of ensuring space is used sustainably. For instance, space resource use might help reduce the environmental pressures caused by human activity on Earth (Cooper, 2003). Yet, the concept of sustainability in outer space remains poorly defined, and scholars diverge in their treatment of this issue (Iliopoulos & Esteban, 2020). Recent efforts have proposed integrating sustainability into space operations through grassroots initiatives (Newman & Williamson, 2018) and soft law instruments (Chrysaki, 2020), aiming to compensate for the absence of binding international regulations.

The concept of harvesting resources from celestial bodies has fascinated people for over a century, first appearing in 19th-century science fiction. Yet it wasn't until the mid-twentieth century, during the Cold War and the space race, that space mining became a topic of serious scientific and policy discourse. The Apollo missions of the 1960s and 1970s, which returned lunar samples, provided early insights into the Moon's geology and sparked interest in its resource potential. By the 1980s and 1990s, growing attention to asteroid composition led scientists to begin categorising near-Earth objects (NEOs) based on their mineralogical value, particularly focusing on water ice and rare metals. Engineers and futurists during this era floated ambitious ideas of asteroid mining to sustain space colonies or offset Earth's resource constraints. However, such visions remained speculative due to immense technological and financial hurdles. The early twenty-first century, however, marked a turning point. Breakthroughs in robotics, miniaturised sensors, and launch capabilities—coupled with the emergence of private aerospace firms like SpaceX, Planetary

Resources, and Deep Space Industries—revived serious interest in space mining. This resurgence coincided with a broader technological shift towards automation and artificial intelligence (AI) in both terrestrial and space applications. Increasingly autonomous planetary rovers, such as those deployed by NASA on Mars, showcased the practicality of operating intelligent machines in remote and hostile environments. Over the past decade, AI has become integral to space exploration, particularly in areas like navigation, terrain analysis, and real-time decision-making. This evolution has significantly advanced the plausibility of AI-driven space mining. Today, both national agencies and commercial entities are actively investigating how machine learning, computer vision, and autonomous robotics can be used not just to identify extraterrestrial resources but to extract and process them directly on-site. The historical arc of space mining thus illustrates a shift from imaginative speculation to a technology-driven reality, increasingly shaped by the capabilities of artificial intelligence.

Space mining stands at the edge of a transformative era in resource extraction, offering the possibility of accessing immense, untapped materials from beyond Earth. However, this potential is accompanied by a host of complex scientific, economic, legal, and ethical challenges. This chapter delves into the diverse aspects of space mining, beginning with the technical obstacles associated with prospecting in remote, data-limited environments and the inherent uncertainties in valuing off-world resources. It explores contentious issues of ownership and governance, highlighting the legal ambiguities and commercial risks posed by the lack of comprehensive international regulatory structures. The chapter provides a critical analysis of current legal frameworks and their limitations in addressing the geopolitical and corporate interests that are increasingly influencing the field. Additionally, it considers the development of effective extraction technologies and the need for frameworks to evaluate and mitigate environmental impacts on celestial bodies. The chapter concludes by presenting a strategic roadmap for integrating artificial intelligence into space mining operations, demonstrating how AI can enable autonomous exploration, improve efficiency, and reduce the risks associated with operating in the extreme conditions of outer space.

2 Space Mining as a New Frontier for Mining

Space mining represents a significant transformation in the history of resource extraction by expanding mining activities beyond Earth and into outer space. While mining has traditionally been focused on Earth's geology, it has long faced challenges such as resource exhaustion, environmental harm, and geopolitical conflicts over mineral-rich areas. As these challenges grow more acute on our planet, celestial bodies like asteroids, the Moon, and Mars are increasingly seen as promising new sources for mining. These extraterrestrial locations are thought to harbour large reserves of valuable materials—including water ice, rare earth metals, platinum-group elements, and helium-3—that are essential for advanced manufacturing and future space development. However, unlike conventional mining on Earth, space mining presents unique technical, financial, and regulatory obstacles. The lack of atmosphere, reduced gravity, and extreme environmental conditions demand entirely new extraction technologies, often relying on autonomous machines and artificial intelligence.

Additionally, the challenge of transporting extracted resources—either back to Earth or for use in space—raises complex engineering and economic issues. Space logistics is an emerging discipline focused on ensuring the effective execution of space missions (Evans, 2006). It encompasses the planning and management of design processes, materials, services, and information essential for mission success (Baraniecka, 2019). It also involves the transportation, storage, and tracking of crew members and equipment required for space operations (Shull et al., 2006). In essence, space logistics supports the movement of materials and personnel beyond Earth's atmosphere, including both outbound missions and the safe return of crews to Earth (Shull et al., 2006). While some definitions exclude Earth-based activities—such as spacecraft design, production, assembly, load handling, and storage—these preparatory steps on Earth are equally vital to mission success (NASA, Bryce Space and Technology, 2024).

For the mining sector, space represents more than just a new physical territory—it signifies a conceptual shift: moving from extraction confined by Earth's limits towards a speculative future where mining transcends our planet. In this sense, space mining both continues traditional extractive practices and radically redefines what mining can be (Bastmeijer & Roura, 2008; Tsing, 2005). It creates opportunities for economic growth, profit, and technological breakthroughs, while simultaneously raising urgent concerns about sustainability, governance, and ownership in environments that have never before been exploited by humans (Wrench, 2019). Consequently, it calls for a fundamental reconsideration of what mining entails, where it can take place, and how its impacts should be managed amid growing planetary and interplanetary interconnectedness.

3 Challenges with Prospecting

Even terrestrial mining is riddled with inefficiencies in the licencing process, characterised by ambiguous procedures, excessive discretionary power, and weak oversight—all of which make the system vulnerable to corruption and manipulation. Information silos across mining departments further hinder coordination, often resulting in geospatial overlaps, conflicting claims, and costly legal disputes. The manual nature of title documentation introduces human error, poor mapping, and inadequate verification processes. In many developing countries, paper-based systems allow for title tampering and conflicting government and individual claims (Hull, 2012). These systemic issues have prompted calls for blockchain adoption, with proposals including digitisation, title tokenisation, and decentralised cadastre systems to enhance transparency and security (Konashevych, 2020).

The spatialisation of inequality in extractive industries has long been recognised, with scholars noting how mining reinforces territorial control and benefits distant power centres while marginalising local communities (Ferguson, 2006; Watts, 2004). The politics of mapping and data infrastructure—especially under digital regimes—further consolidate power through spatial knowledge and bureaucratic control. These terrestrial challenges are highly relevant to space mining, Table 1. As humanity expands extractive frontiers beyond Earth, the risks of repeating and amplifying terrestrial governance failures—through opaque title claims, inadequate mapping, and inequitable access—remain high. The adaptation of blockchain and decentralised infrastructure in space mining could pre-emptively address such issues, ensuring more equitable, transparent, and verifiable systems of resource ownership and management in extra-planetary environments. Exploring minerals in space involves a fundamentally different set of challenges compared to terrestrial mining. Unlike Earth, where prospecting benefits from extensive geological surveys, satellite imagery, and decades of mining experience, space exploration faces radical uncertainty. Most celestial bodies are still poorly mapped, and their composition is often estimated indirectly through methods like spectral analysis, remote sensing, or comparisons with meteorites. This scarcity of direct data complicates assessments of not only the presence of valuable resources but also their concentration, accessibility, and extraction potential. Distance and communication delays add further complexity. Missions targeting near-Earth asteroids or the Moon require years of preparation and depend heavily on autonomous systems that must traverse rugged terrain with minimal real-time guidance from Earth. Instruments need to detect subsurface materials without the support of on-site geological teams or heavy drilling equipment. Although robotic landers and orbiters provide valuable data, they are limited by resolution, mobility, and durability in harsh space environments. Moreover, space prospecting faces constraints related to launch weight, mission duration, and power supply. Instruments must be compact and energy-efficient, often compromising detail or depth for practicality. These factors make it challenging to conduct the thorough, multi-stage evaluations typical of Earth-based exploration. Consequently, early space mining efforts often proceed based on incomplete or probabilistic geological data, increasing the risk that costly missions may yield little return. The absence of standardised systems for classifying and valuing space resources further complicates coordination among stakeholders. Without common frameworks for categorising celestial deposits or assessing their commercial worth, prospecting activities remain fragmented and speculative. Addressing these challenges will demand significant progress in autonomous sensing technology, advanced long-distance data modelling, and international cooperation through open-access space data platforms.

Table 1 Terrestrial challenges

Challenge category	Ordinary mining	Space mining
Environmental conditions	Operates in known terrestrial environments with atmospheric protection	Harsh, vacuum or near-vacuum environments with extreme temperatures and radiation exposure
Geological data	Extensive geological surveys, mapping, and historical mining data available	Limited, indirect data from remote sensing and meteorite analogues; most bodies are poorly mapped
Resource accessibility	Relatively accessible deposits; infrastructure supports drilling and excavation	Difficult to access due to distance, low gravity, unknown terrain, and lack of established infrastructure
Technology requirements	Uses well-established mining and processing technology	Requires novel autonomous robotics, AI, and new extraction methods adapted to zero or low gravity
Transportation logistics	Established supply chains and transport networks	High launch costs, limited payload capacity, propulsion constraints, and long travel times
Regulatory frameworks	Clear national and international mining laws and environmental regulations	Legal ambiguity, lack of clear ownership and resource extraction laws, and weak enforcement
Environmental impact	Monitored environmental impact assessments and remediation protocols	Uncertain environmental effects with no established impact assessment or protection frameworks
Operational control	Direct human oversight is possible	Heavy reliance on autonomous or semi-autonomous systems due to communication delays
Safety and risk management	Established protocols for worker safety and operational risk	High technical risk, remote operations, limited repair options, and high mission failure costs
Market and valuation	Mature commodity markets with transparent pricing	Speculative markets, uncertain resource value, and lack of standard pricing frameworks
Ownership and disputes	Well-defined property rights and dispute resolution mechanisms	Ambiguous ownership laws, potential for territorial disputes, and regulatory arbitrage
Cost structure	Costs spread across labour, equipment, energy, and regulatory compliance	Massive upfront investment, high launch and operational costs, and expensive technology development

4 Challenges with Valuations

Assigning value to extraterrestrial resources involves both conceptual and practical challenges, as traditional economic models are ill-suited to the distinct conditions of space. Unlike Earth-based mining—where mineral prices are shaped by established markets, infrastructure, and geopolitical dynamics—space mining exists in a speculative, largely unregulated frontier. With no active marketplace for materials like lunar water ice or asteroid platinum, valuations depend heavily on hypothetical demand, projected scarcity, and uncertain technological viability. A key difficulty lies in contextualising value: resources highly prized on Earth may not be as valuable in space, and vice versa. For instance, platinum-group metals command high prices on Earth, but the costs of extracting and transporting them from space could outweigh their market worth. Conversely, water ice, which has limited value on Earth, could be extremely valuable in space, serving as life support or rocket fuel and potentially reducing reliance on expensive Earth launches. Estimating costs adds further complexity. The true value of a space mining venture must account not only for the market price of resources but also the extraordinary expenses involved in prospecting, extraction, transport, and risk management—costs that remain difficult to quantify due to the developmental status of most space mining technologies. Without precedents, investors and policymakers rely on speculative projections with significant uncertainty. Legal and ownership ambiguities also cloud valuation. Since property rights over celestial resources remain unsettled under treaties like the Outer Space Treaty, assigning concrete monetary value is speculative and may discourage investment. Valuations could vary widely depending on future legal frameworks: resources secured under clear international ownership would carry greater financial weight than those subject to ambiguous or contested claims. Ultimately, valuing space resources transcends technical and economic considerations; it involves forecasting future regulatory, technological, and logistical realities. Until clearer legal regimes and more concrete technological benchmarks emerge, valuations will remain speculative, shaped by varying degrees of risk tolerance and visionary thinking.

Stewart Myers (1977) from MIT Sloan School of Management introduced the concept that a firm's market value consists not only of the discounted value of expected cash flows but also of its "growth opportunities," particularly when these involve tangible assets. These were later termed "real options" (RO), whose value is captured through "Real Options Value" (ROV), and evaluated using "Real Options Analysis" (ROA). ROA parallels financial option valuation, extending traditional Discounted Cash Flow (DCF) analysis by incorporating managerial flexibility in response to unfolding uncertainties—something standard static Net Present Value (NPV) fails to do.

In mining projects, such flexibility might be embedded (e.g., adjusting grades vs. tonnage) or created (e.g., through contract design), with real options ranging from simple (e.g., defer investment) to complex combinations (e.g., adjust output based on commodity prices or mothball operations temporarily) (Guj & Chandra, 2019). Trigeorgis (1995) provides a detailed review of ROA methods used in the resource sector, citing foundational studies by Tourinho (1979), McDonald and Siegel (1986), and Paddock et al. (1988), and later contributions by Copeland and Antikarov (2001) and Dixit and Pindyck (1994). Savolainen (2016) offers a mining-specific review of ROA applications, including Brennan and Schwartz's (1985) evaluation of Chilean copper mines, and provides comparative assessments of methodologies such as Black–Scholes (BSF), binomial lattices, and Monte Carlo simulations. BSF-based methods, which treat uncertainty as continuous-time stochastic processes, are analytically elegant but pose challenges when applied to illiquid, non-traded assets like mining projects, where reliable market data is scarce. As Copeland and Antikarov (2001) note, this limits accurate estimation of key inputs like volatility (σ), spot value (S), and exercise price (X). To address this, the Marketed Asset Disclaimer (MAD) approach proposes using a project's DCF-calculated NPV as a proxy for spot value (S), capital cost as the exercise price (X), and Monte Carlo simulations to estimate volatility (Copeland & Antikarov, 2001). However, aggregating multiple sources of uncertainty into a single volatility measure often introduces upward bias. This complexity becomes evident in evaluating options like purchasing a mining project outright versus acquiring a two-year purchase option. Classical ROA methods such as Black–Scholes and binomial models rely on aggregated volatility, while Monte Carlo and decision tree methods allow for disaggregated inputs, offering greater nuance. Yet, even these are hampered by the lack of observable market data for mining assets, making volatility estimation and value derivation error-prone (Copeland & Antikarov, 2001). These insights are directly relevant to space mining, a domain similarly characterised by high uncertainty, long lead times, and limited market comparables. Like terrestrial mining, space mining projects face uncertain yields (e.g., asteroid composition), volatile potential returns (e.g., due to future commodity prices on Earth or in space), and capital-intensive development. Applying ROA in this context allows for valuation that captures managerial flexibility in decision-making—whether to delay, proceed, expand, or abandon based on technological readiness or regulatory shifts. However, just as in terrestrial projects, space mining ventures suffer from a lack of historical price data and traded benchmarks. The MAD approach could theoretically help simulate the cash flows and volatility needed for ROA, though with acknowledged limitations. As in the mining sector, Monte Carlo simulation and decision tree analyses may offer more accurate valuation tools than standard Black–Scholes methods due to their ability to accommodate disaggregated, project-specific uncertainties. Ultimately, ROA provides a structured way to assess strategic flexibility in space mining investment decisions. However, its reliability hinges on careful modelling of risks unique to extraterrestrial operations, and we will shortly consider how the incorporation of Artificial Intelligence in space mining can help with precisely this limitation.

5 Challenges with Ownership

Mining properties, including mineral property, unified estate, severed or split estate, and fractional ownership, often lead to disputes over surface and subsurface ownership. For example, in a unified estate, both surface and mining rights are owned together, whereas, in severed estates, the ownership is split, leading to potential conflicts between landowners and mining companies (Kent, 2019). Technological advancements have expanded extraction processes, causing new conflicts, particularly in split estate situations where mineral rights dominate surface rights (Ryder & Hall, 2017). In the US, the Stock-Raising Homestead Act of 1916 created split estates, exacerbating issues between surface and mineral rights owners (Collins & Nkansah, 2015; Libecap, 1978). Mineral rights are also contentious due to the rule of capture, allowing extraction even from beneath another's property.

The question of who owns resources mined in space lies at the core of legal and geopolitical uncertainty in the nascent space economy. Unlike terrestrial mining, which is governed by established national laws and international agreements, space mining exists in a legal grey area shaped by Cold War–era treaties not designed to address commercial extraction. The 1967 Outer Space Treaty (OST), ratified by over 110 countries—including key spacefaring nations—declares outer space, including the Moon and other celestial bodies, as the "province of all mankind," prohibiting national sovereignty claims over these territories. However, the treaty does not explicitly address the rights of private or non-state actors to extract and own resources from space. This ambiguity has led to varied national interpretations. Countries like the United States, Luxembourg, the United Arab Emirates, and Japan have enacted laws granting their citizens or corporations rights to extract and possess extraterrestrial resources. These legislative moves aim to encourage investment and innovation by reducing legal uncertainties but also raise tensions with the spirit, if not the letter, of international space law. Critics warn that such unilateral actions risk enabling de facto territorial claims, undermining the principle of space as a shared global commons.

Another ownership challenge involves potential conflicts between state and corporate interests, especially as private entities increasingly participate in space exploration. In the absence of clear multilateral agreements, disputes over resource claims, extraction zones, or overlapping missions could spark legal conflicts or geopolitical tensions. The lack of effective enforcement mechanisms in current treaties worsens these risks. Moreover, the concept of ownership in space raises ethical and distributive justice issues. If access to space resources becomes concentrated among a few wealthy states and corporations, it could exacerbate global inequalities and replicate patterns reminiscent of historical extractive colonialism. Proposals for more equitable governance frameworks—such as benefit-sharing arrangements or the creation of a new international regulatory body—have yet to gain significant traction in formal negotiations. Thus, the question of ownership in space mining is not merely a legal technicality but a fundamental issue that will shape the future of space governance.

Without more explicit, enforceable, and internationally recognised rules, ownership disputes will persist, potentially threatening the legitimacy and sustainability of extraterrestrial resource extraction.

6 Challenges with Logistics

As space logistics involves planning, implementing, and controlling missions from low Earth orbit (LEO) to deep space or beyond the atmosphere (Lee et al., 2008), the process involves the production of materials to be used in space missions, equipment providing mobility in space, essential life support materials, spare parts, and ensuring the successful return of the crew to the starting point. Space logistics is the sum of activities such as supply, transportation, handling, storage, and service provision to ensure the sustainability of activities in space (Gralla & de Weck, 2012). The increasing interest in space exploration is expected to lead to the development of satellites, space stations, space transportation, and space exploration vehicles, as well as the commercialisation of space (Kulu, 2021). Additionally, planning for activities such as space tourism, energy production, and mining in space supports this situation (Wilkinson, 2015). The most recent example is the announcement by Yuri Borisov, the head of Russia's space agency Roscosmos, that Russia and China will build a nuclear power plant on the Moon between 2033 and 2035 (Reuters, 2024).

Furthermore, in the concept of Industry 4.0 or smart production, efficient and effective production will be ensured using technologies such as artificial intelligence, big data, Internet of Things (IoT), machine learning, cloud computing, etc. (Eugeni et al., 2022). The use of these technologies involves the use of satellites, one of the elements of space logistics. Data transfer will be facilitated more easily and effectively through satellites. These technologies are important for supporting the space economy or industry (ibid.). The use of these technologies can provide advantages in terms of speed, time, and efficiency through satellite-based internet services. Statistical data shows that investments in the space industry have increased since the 2000s. Especially between 2017 and 2022, the space economy has reached a significant size. Space offers commercial opportunities. However, to ensure the sustainability of logistic activities in space, space policy and infrastructure need to be developed (Evans, 2006). The increasing interest in space activities will lead to rapid commercialisation of space in the future, which will make space logistics, and therefore carrier rockets and vehicles, more important. Possible collisions pose a risk for satellites and spacecraft. Hence, space debris is one of the important issues in space logistics (Liou et al., 2004). Environmental risks refer to the impact of the natural environment on the spacecraft before, during, and after the transportation process on Earth and in space (Kirov et al., 2022).

The logistical challenges of space mining rank among the most daunting obstacles to its practical implementation. Unlike mining on Earth—where supply chains, labour, equipment, and energy are relatively accessible—space mining must overcome vast distances, hostile environments, and severe constraints on transport and infrastructure. Every stage of the process—from launch and travel to landing, extraction, processing, and either utilisation in space or transport back to Earth—demands cutting-edge technologies, precise execution, and substantial upfront investment. A central logistical hurdle is transportation. Sending spacecraft, robotic mining equipment, and extraction systems to bodies like asteroids, the Moon, or Mars is limited by payload capacity, fuel efficiency, and high launch costs. Returning mined materials to Earth presents even greater engineering and financial challenges. Existing propulsion methods restrict the speed and scale of such missions, while refuelling options in deep space remain largely theoretical. Operational autonomy is another key challenge. Communication delays ranging from seconds to over twenty minutes—depending on distance—make real-time control from Earth impractical. As a result, space mining operations must rely heavily on autonomous or semi-autonomous systems capable of navigation, decision-making, extraction, and maintenance without continuous human intervention. This demands advanced integration of artificial intelligence, robotics, and machine learning—fields that are still rapidly evolving.

Infrastructure limitations further restrict space operations. Celestial bodies lack atmospheric shielding, roads, power grids, or built environments to support machinery. Equipment must function in vacuum or near-vacuum conditions, endure abrasive dust, and tolerate extreme temperature swings. Even basic tasks like anchoring equipment on a low-gravity asteroid or storing mined materials require innovative engineering solutions. Lastly, long-term mission planning and risk management remain in early stages. Space mining carries enormous uncertainties: technical failures, unforeseen surface conditions, and inaccurate resource estimates can quickly jeopardise a mission. The remote nature of these operations means repairs or recovery options are extremely limited, heightening the consequences of logistical errors. Overall, the logistical challenges of space mining go beyond simply adapting terrestrial mining methods—they necessitate a complete rethinking of extraction, transportation, and operational systems tailored to entirely alien environments.

7 Space Mining and the Possibility of Irregularities

In general, even terrestrial mining suffers from the following problems of irregularities. There is no reason why these problems will not exist in space mining. When ownership is not fixed, there is going to be an even greater possibility of contestations, this time between countries vying for greater control over space mines.

The disjointed data repositories were riddled with numerous errors and inconsistencies, including administrative data, spelling errors, duplicated records, multiple copies of the same database with different content, and absent data. These errors are likely to multiply if space mining is not ready for the rapid adoption of artificial intelligence, Fig. 1. For geometrical or graphic data, plotting inaccuracies result in overlaps, multiple copies of the same concession in the same or different areas, insufficient links to administrative data, or the same database on different unconnected computers. The result was the need to go through a long and expensive process of data cleaning, consolidation, and migration to the new computerised platform. Nyaga et al. (2020) identified legal hurdles, lack of technical expertise in record keeping, and inadequate funding as serious challenges affecting the implementation of the cadastre system at the Ministry of Mining Headquarters in Nairobi. A key concern in space mining is regulatory arbitrage, where companies or nations exploit gaps and inconsistencies in national and international space law to make ambiguous claims or evade oversight. With no universally accepted system for licencing, inspection, or resource reporting, compliance could be uneven and monitoring complicated. This opens the door to false claims about resource yields, undisclosed extraction activities, and hidden environmental impacts. Corporate secrecy and intellectual property protections add to the opacity, as companies compete to maintain technological advantages by withholding crucial information on prospecting methods, extraction techniques, or mining locations. Such a lack of transparency could hinder collective learning, heighten geopolitical tensions, and obstruct the establishment of shared best practices. Irregularities might also emerge from informal or extra-legal claims on celestial bodies. As multiple actors operate in overlapping or contested areas—especially on the Moon or among valuable near-Earth asteroids—disputes could intensify without clear arbitration or conflict resolution mechanisms. In the absence of enforceable norms, dominant players might assert control through technological superiority or political influence, effectively privatising common space resources. Environmental and ethical irregularities also pose risks. Mining could damage celestial bodies without prior environmental assessments or destroy scientifically significant sites in pursuit of profit.

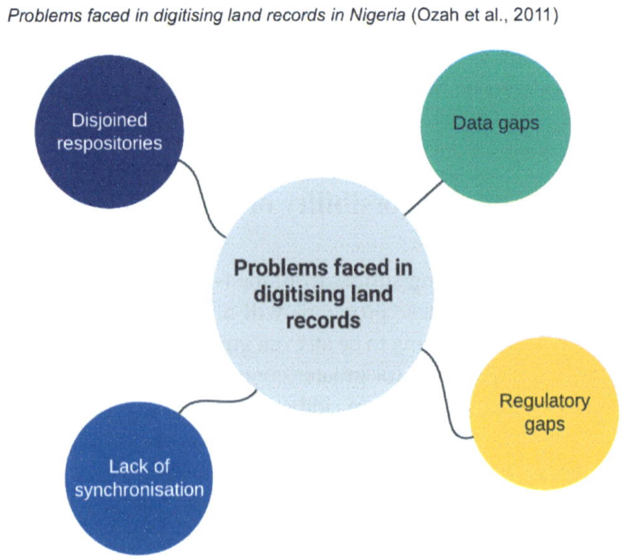

Fig. 1 Problem faced in digitising land records

Since current treaties lack environmental protections and enforcement mechanisms beyond Earth, such activities could proceed without meaningful accountability. Valuation of space-derived minerals is another area prone to irregularities. Without established markets or pricing systems for extraterrestrial resources, companies might inflate or inconsistently report valuations to attract investment or shape public opinion. The speculative nature of the space economy makes it vulnerable to hype cycles, market distortions, and uneven investor risk exposure, potentially leading to financial instability or deliberate misrepresentation in the absence of standardised verification protocols. Finally, the sector is exposed to financial speculation and hype-driven investment cycles. Inflated valuations, unrealistic projections, and limited public oversight may generate economic bubbles, misallocated capital, or fraudulent schemes masked as innovation. Without a strong, enforceable governance framework, space mining risks replicating and amplifying many irregularities historically seen in unregulated extractive frontiers on Earth. Addressing these challenges will require international cooperation, legal innovation, and increased public scrutiny to ensure a sustainable and transparent space mining industry.

8 International Regulatory Regimes

The emerging field of space mining operates within a complex and fragmented mix of international treaties and national laws that were not originally crafted to regulate the commercial extraction of extraterrestrial resources. This legal uncertainty creates opportunities for states and corporations to exploit these frameworks, advancing their interests with minimal oversight. Foundational agreements like the 1967 Outer Space Treaty (OST) and the 1979 Moon Agreement set broad principles—such as the non-appropriation of outer space and the peaceful use of celestial bodies—but lack specific provisions to govern resource ownership, extraction rights, or benefit-sharing. In response to these gaps, some countries and companies have enacted domestic laws granting their citizens the right to mine and possess space resources, effectively sidestepping the intended spirit of international agreements. Such unilateral measures risk fragmenting the regulatory environment, encouraging conflicting claims and "forum shopping," where actors register in jurisdictions with lax regulations to avoid stricter oversight elsewhere.

Furthermore, the absence of a centralised enforcement body means compliance with international norms depends primarily on voluntary cooperation, diplomatic negotiations, or geopolitical power dynamics. This weak enforcement allows influential players to test the limits of permissible activity, undermining collective governance efforts. The potential for regulatory arbitrage also increases the danger of a "race to the bottom," where competitive advantage is prioritised over ethical or environmental responsibilities. Exploiting these international regimes may deepen existing disparities between spacefaring nations and those lacking space capabilities, as well as between private corporations and the broader global community. Without inclusive, transparent, and equitable governance frameworks, space mining risks repeating patterns of resource exploitation and governance failures familiar from terrestrial extractive industries—benefiting a small elite while marginalising others. To address these challenges, urgent multilateral dialogue is needed to update and harmonise international space law, establish clear norms for resource extraction, and develop mechanisms for dispute resolution and benefit-sharing. Such reforms are crucial to preventing the exploitation of legal gaps and fostering a sustainable, just space mining industry.

9 Assessing Environmental Damage

As space mining moves from concept to reality, evaluating its potential environmental impacts on celestial bodies emerges as a critical yet highly complex challenge. Unlike Earth—where environmental assessments are supported by extensive ecological data and established regulatory systems—space environments remain largely unexplored and poorly understood. This lack of baseline knowledge complicates attempts to predict and measure the effects of mining on fragile extraterrestrial ecosystems, unique geological formations, and scientifically valuable heritage sites. A fundamental challenge is defining what counts as "environmental damage" in space. While bodies like asteroids, the Moon, and Mars do not host known biospheres, they contain distinct geological features and regolith layers that preserve vital records of planetary history and solar system evolution. Mining activities such as excavation, blasting, or waste disposal risk destroying this irreplaceable scientific information.

Furthermore, the long-term consequences of mining-generated debris, dust clouds, or contamination on local space environments and orbital paths are still poorly understood. The lack of comprehensive environmental governance frameworks further deepens this uncertainty. Existing international treaties do not explicitly require environmental impact assessments for space activities, nor are there standardised protocols for monitoring or mitigating environmental harm. This regulatory gap risks mining operations proceeding without adequate safeguards, potentially causing irreversible changes to extraterrestrial landscapes. Technological constraints also limit assessment capabilities. Remote sensing and robotic explorers currently provide only partial, low-resolution data, making it difficult to fully assess the extent of environmental disruption. Effective monitoring may require continuous, real-time data collection—necessitating substantial investment in autonomous sensing and communication infrastructure. Given these complexities, developing environmental impact assessment protocols explicitly tailored to space mining is urgently needed. Such frameworks should draw on multidisciplinary expertise—from planetary science and ecology to ethics and international law—and incorporate precautionary principles to protect celestial environments. Integrating environmental considerations early on will be essential to balancing the economic ambitions of space mining with the preservation of humanity's shared extraterrestrial heritage.

Table 2 provides a risk-assessment protocol that may be used to detect environmental damages as well as infringements of regulations through the incorporation of Artificial intelligence in the processes of space mining detailed above.

10 Implementation of Artificial Intelligence: A Road Map

The factors above necessitate examining the possibilities and challenges of blockchain implementation in managing these disputes. Partial digitisation is time-consuming and requires specific training for technical and non-technical employees (Deshpande, 2019). Knowledge sharing can enhance digital innovation (Malhotra et al., 2005), but proper organisational adaptation is essential for successful innovation.

Blockchain technology, which can revolutionise space mining by providing immutable, decentralised records, offers significant potential. It can also be used to assess ownership on a rotating basis; space minerals are considered global property. Abramov et al. (2021) argue for using innovative technologies in geoinformation management. Blockchain could address issues in mining land management by linking each transaction to a previous digital signature, ensuring data integrity and transparency. This technology provides secure, long-term data storage and establishes accountability. However, successful blockchain implementation requires substantial changes in business processes and the commitment of involved parties. The complexity of mining property ownership necessitates innovative solutions like blockchain to address conflicts and enhance transparency. While digitisation poses challenges, its potential benefits for governance and land management are substantial. Effective implementation of blockchain can lead to improved dispute resolution and better management of resources in the mining sector. All of the following processes need to be centrally managed by dynamic algorithms that can account for rotational ownership, detect information asymmetries, identify logistical challenges, and detect violations in regulations.

Blockchains, for instance, can create greater transparency and immutability and help avoid conflicts and contestations that terrestrial mining companies have faced. Blockchain is a growing list of records or blocks linked using cryptography (Morris, 2016). Each block contains a cryptographic hash of the previous block, reinforcing the integrity of the chain. Managed in a peer-to-peer network, blockchains serve as distributed ledgers where nodes validate new blocks through a collective protocol. This design ensures that data in any block cannot be retroactively altered without altering all subsequent blocks, providing secure and reliable data storage. This feature is particularly useful for land title management, which often struggles with data integrity in manual storage systems. In the digital space, identities are constantly negotiated and shaped by digital identity, which influences perceptions of physical identity (Marwick, 2013). Digital identity can liberate individuals from physical distinctions such as race, class, and gender (Turkle, 1995). Self-sovereign identity (SSI) offers a solution by giving individuals autonomous control over their digital

Table 2 Risk-assessment protocol that may be used to detect environmental damages

Risk assessment processes	
Risk assessment	
Analyse and evaluate risks	Tools like ATHENA can be utilised to assess and evaluate risks, taking into account reputational risks that can affect stakeholder trust and the long-term viability of mining operations
Risk reporting	
Communicate threats and opportunities	Effective risk reporting involves communicating potential threats and opportunities arising from corporate activities and external factors, including their impact on community relations and crisis management capabilities
Decision	
Informed risk decisions	Making informed choices regarding whether to accept, avoid, transfer, or mitigate risks based on their potential impact on the sector's objectives (Johnson, 2021)
Risk treatment	
Implement control measures	Control measures, such as preventive, corrective, directive, and detective controls, are essential to manage risks effectively
Residual risk reporting	
Document and report residual risks	It's crucial to document and report risks that remain after treatment measures have been applied
Ongoing monitoring	
Continuous monitoring	Continuously monitoring the risk environment ensures effective control and helps identify new risks as they emerge
Risk identification and treatment	
Reputational risk management	In Zimbabwe's mining sector, factors such as financial performance, leadership, regulatory compliance, delivery of promises to customers, workplace talent and culture, CSR, and crisis communication are critical areas for diligent risk identification and management (Jones, 2018)
Control types	
Preventive controls	Robust cybersecurity measures to prevent data breaches
Corrective controls	Restoration and compensation measures for environmental damage
Directive controls	Compliance protocols to prevent regulatory infractions

(continued)

Table 2 (continued)

Risk assessment processes	
Detective controls	Audit trails in financial systems to detect anomalies
Risk analysis tools and regulatory references	
Use of risk analysis tools	Tools like ATHENA enhance decision-making by providing insights into potential risks and their impacts
Global regulatory frameworks	Frameworks such as the EU AI Act and US White House Executive Order on AI provide guidelines for the responsible use of AI and related technologies, offering a blueprint for Zimbabwe to formulate its own AI governance framework (Johnson, 2021)
Comprehensive approach to risk management	
Balancing technological advancements and risk management	Incorporating AIS can help govern AI applications and augment human decision-making in the mining sector, ensuring that operations are efficient and compliant with emerging international standards (Jones, 2018)

identities, ensuring anonymity while establishing trust. SSI operates by allowing one party to provide credentials to another, with the recipient trusting the credential provider, thus transferring trust to the credential holder.

10.1 Information Asymmetries or Inefficiencies

Artificial intelligence (AI) holds transformative potential to tackle many of the multifaceted challenges confronting space mining, enabling more efficient, autonomous, and adaptive operations in the harsh and remote extraterrestrial environment, Fig. 2. By integrating AI-driven technologies, space mining ventures can significantly reduce risks, optimise resource utilisation, and enhance decision-making across all phases—from prospecting and valuation to extraction and logistics. AI-powered remote sensing and machine learning algorithms can analyse vast datasets from telescopes, satellites, and robotic probes to improve the identification and characterisation of mineral-rich celestial bodies. Through pattern recognition and predictive modelling, AI can compensate for limited direct data, enabling more precise estimations of resource location, concentration, and extractability. By integrating economic modelling with real-time geological data, AI systems can dynamically update valuations of space resources. Advanced simulation tools can forecast extraction costs, market fluctuations, and technological advancements, allowing stakeholders to make informed investment decisions despite inherent uncertainties. AI can assist in tracking

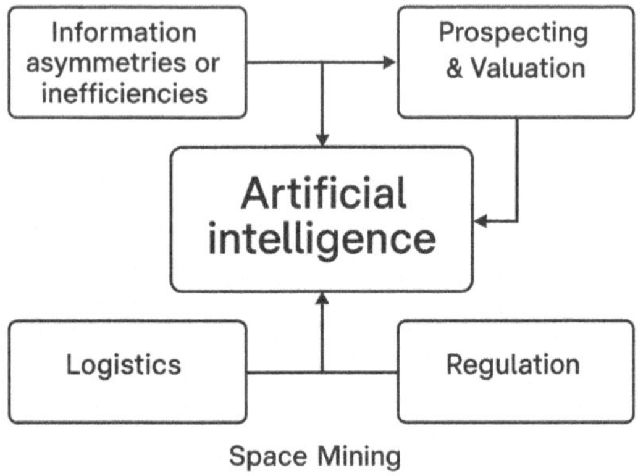

Fig. 2 The role of Artificial Intelligence in space mining

claims, licences, and compliance with evolving international and national regulations by automating documentation, monitoring activity, and flagging potential legal conflicts. Intelligent systems can also facilitate transparent reporting and data sharing, promoting accountability and minimising disputes. Autonomous robotic systems powered by AI can perform complex tasks such as landing, drilling, extraction, and material handling with minimal human intervention. Machine learning enables these robots to adapt to unexpected terrain features or equipment malfunctions, increasing mission resilience. AI-driven mission planning optimises trajectories, fuel consumption, and timing, reducing costs and enhancing safety. By providing real-time monitoring and data validation, AI can detect anomalies, prevent fraudulent claims, and improve transparency in resource reporting and environmental impact assessments. Blockchain integrated with AI can offer secure, immutable records of mining activities, fostering trust among stakeholders. AI-powered sensors and data analytics enable continuous environmental monitoring, detecting changes in surface morphology, dust dispersion, or contamination. Predictive models can simulate the ecological consequences of mining actions, guiding operators to minimise harm and comply with environmental protocols.

AI excels at processing incomplete or noisy data and can assist human operators and policymakers by providing probabilistic risk assessments and scenario analyses. This capacity is critical in the inherently uncertain context of space mining.

10.2 Logistics

Logistics in space mining involve coordinating the transport, operation, and maintenance of mining equipment across vast distances and hostile environments—tasks that are uniquely difficult due to communication delays, limited infrastructure, and extreme conditions. AI can revolutionise these logistics by enabling autonomous, adaptive, and efficient management of operations, reducing reliance on constant human intervention. AI-powered autonomous systems can pilot spacecraft and robotic miners with high precision, navigating unpredictable terrains on asteroids or the Moon without direct human control. Machine learning algorithms enable these systems to learn from environmental feedback and adjust their actions in real time, improving safety and mission success. AI algorithms can monitor equipment health continuously by analysing sensor data to predict failures before they occur. This predictive maintenance capability reduces downtime and the risks associated with unexpected breakdowns in remote, inaccessible locations. AI can analyse vast amounts of data to optimise launch windows, travel routes, energy consumption, and resource allocation. By simulating different logistical scenarios, AI helps minimise costs and maximise operational efficiency in complex, multi-phase missions. In a space mining context, AI can manage limited resources like power, fuel, and spare parts, allocating them efficiently according to real-time needs and mission priorities. This ensures the sustainability of long-duration missions with minimal waste. AI can prioritise and manage communication data streams between Earth and mining operations, compensating for signal delays and bandwidth constraints. It can autonomously decide which data are critical to transmit immediately versus which can be processed onboard.

10.3 Regulation

AI-powered satellites and ground stations can continuously monitor space activities—such as mining operations, spacecraft movements, and resource extraction—by analysing telemetry, imagery, and communication signals. Machine learning algorithms can identify anomalies or unauthorised activities that may signal breaches of treaties like the Outer Space Treaty or national licencing conditions. By training on historical data and known compliance behaviours, AI systems can detect unusual patterns that deviate from legal or operational norms. For example, unexpected extraction activity in restricted zones, unauthorised debris generation, or suspicious alteration of orbital paths can trigger alerts for further investigation. Combining AI with blockchain technologies enables the creation of immutable, transparent records of mining operations, licences, and resource transfers. AI can automatically verify transactions and activities against regulatory requirements, flagging discrepancies or potential fraud. AI systems can aggregate and analyse data from multiple national and private sensors, supporting collaborative monitoring efforts among spacefaring nations. Shared AI platforms can help standardise compliance reporting and enhance trust among stakeholders. By providing timely and accurate information on violations, AI assists regulatory bodies in enforcing conventions more effectively. It enables proactive responses rather than reactive investigations, helping to prevent conflicts or environmental harm before they escalate.

11 Conclusion

Space mining stands at the threshold of transforming humanity's relationship with natural resources, extending extraction activities beyond Earth and opening unprecedented economic and scientific opportunities. However, this new frontier presents a complex array of challenges—including prospecting uncertainties, valuation difficulties, ownership disputes, logistical hurdles, environmental risks, and the potential for legal and ethical irregularities. Addressing these challenges is essential to ensure that space mining develops in a sustainable, equitable, and accountable manner. Artificial intelligence emerges as a critical enabler in this context. By enhancing data analysis, enabling autonomous operations, optimising logistics, and supporting regulatory compliance, AI can significantly mitigate the risks inherent in space mining ventures. Its capacity for adaptive decision-making, predictive modelling, and real-time monitoring provides the technological backbone needed to operate effectively in the remote and harsh environments of outer space.

Nevertheless, the deployment of AI in space mining must be accompanied by robust international cooperation, transparent governance frameworks, and ethical considerations to prevent misuse and ensure fair benefit-sharing. Only through a multidisciplinary approach that combines technological innovation with legal and policy reforms can space mining fulfil its promise as a new frontier for sustainable resource development. The integration of AI within space mining not only addresses many of its pressing challenges but also paves the way for a future where humanity can responsibly harness the vast resources of the cosmos.

Bibliography

Abramov, V., Sikarev, I., Chusov, A., Tatarnikova, T., & Shilin, M. (2021). Digital transformation of expert systems design within geoinformation management in the Arctic. *Journal of Physics: Conference Series, 2001*, 012014. https://doi.org/10.1088/1742-6596/2001/1/012014

Agrawal, A., Gans, J., & Goldfarb, A. (2019). Exploring the impact of artificial intelligence: Prediction versus judgment. *Information Economics and Policy, 47*. https://doi.org/10.1016/j.infoecopol.2019.05.001

Baraniecka, A. (2019). Space logistics—Current status and perspectives. *Transport Economics and Logistics, 82*, 67–78.

Bastmeijer, K., & Roura, R. (2008). Environmental impact assessment in Antarctica. *Impact Assessment and Project Appraisal, 26*(1), 27–42.

Brennan, M. J., & Schwartz, E. S. (1985). Evaluating natural resource investments. *The Journal of Business, 58*(2), 135–157. Retrieved 18 Nov 2025 from http://www.jstor.org/stable/2352967

Chrysaki, M. (2020). The sustainable commercialisation of space: The case for a voluntary code of conduct for the space industry. *Space Policy, 52*, 101375. https://doi.org/10.1016/j.spacepol.2020.101375

Collins, A. R., & Nkansah, K. (2015). Divided rights, expanded conflict: Split estate impacts on surface owner perceptions of shale gas drilling. *Land Economics, 91*(4), 688–703.

Cooper, L. A. (2003). Encouraging space exploration through a new application of space property rights. *Space Policy, 19*(2), 111–118. https://doi.org/10.1016/S0265-9646(03)00016-X

Copeland, T., & Antikarov, V. (2001). *Real options: A practitioner's guide*. Texere.

Delgado-López, L. (2015). Beyond the Moon Agreement: Norms of responsible behaviour for private sector activities on the Moon and celestial bodies. *Space Policy, 33*(1), 1–3. https://doi.org/10.1016/j.spacepol.2014.08.006

Deshpande, A. (2019). An inquiry into the impact of digitisation and customised ERP applications on Twin Engineers' overall efficiency—An empirical study approach. *International Journal of Business Analytics and Intelligence, 7*(1), 73–87.

Dixit, A. K., & Pindyck, R. S. (1994). *Investment under uncertainty*. Princeton University Press. https://doi.org/10.2307/j.ctt7sncv

Eugeni, M., Quercia, T., Bernabei, M., Boschetto, A., Costantino, F., Lampani, L., & Gaudenzi, P. (2022). An industry 4.0 approach to large-scale production of satellite constellations: The case study of composite sandwich panel manufacturing. *Acta Astronautica, 192*, 276–290.

Evans, W. (2006). *Logistics and supply chain management: A space operations enabler*. In SpaceOps 2006 Conference.

Ferguson, J. (2006). *Global shadows: Africa in the neoliberal world order*. Duke University Press.

Genta, G. (2014). Private space exploration: A new way for starting a spacefaring society? *Acta Astronautica, 104*, 480–486. https://doi.org/10.1016/j.actaastro.2014.04.008

Gralla, E. L., & de Weck, O. (2012). *On-orbit assembly strategies for next-generation space exploration*. In 57th International Astronautical Congress.

Guj, P., & Chandra, A. (2019). Comparing different real option valuation approaches as applied to a copper mine. *Resources Policy, 61*, 180–189.

Hull, M. S. (2012). *Government of paper: The materiality of bureaucracy in urban Pakistan*. University of California Press.

Iliopoulos, N., & Esteban, M. (2020). Sustainable space exploration and its relevance to the privatisation of space ventures. *Acta Astronautica, 167*, 85–92. https://doi.org/10.1016/j.actaastro.2019.09.037

Johnson, T. (2021). Global regulatory frameworks for AI in mining. *International Journal of Mining and Technology, 18*(4), 456–478.

Jones, L. (2018). Reputational risk and its management in the mining industry. *Corporate Governance Journal, 15*(2), 289–305.

Kent, C. A. (2019). Surface and subsurface property rights in minerals: The continuing conflict. *The Journal of Private Enterprise, 34*(2), 43–64.

Kirov, B., Georgieva, K., & Asenovski, S. (2022). *Space weather and its effects on spacecraft charging*. In Fourteenth Workshop.

Konashevych, O. (2020). Cross-blockchain protocol for public registries. *International Journal of Web Information Systems, 16*(4), 32–48. https://doi.org/10.1108/IJWIS-07-2020-0045

Kulu, E. (2021). *In-space economy in 2021—Statistical overview and classification of commercial entities*. In 72nd International Astronautical Congress (IAC), Dubai.

Lee, G., Jordan, E., Shishko, R., de Weck, O., Armar, N., & Siddiqi, A. (2008). *SpaceNet: Modeling and simulating space logistics*. In AIAA SPACE Conference & Exposition.

Libecap, G. D. (1978). Economic variables and the development of the law: The case of western mineral rights. *The Journal of Economic History, 38*(2), 338–362. https://doi.org/10.1017/S0022050700123732

Liou, J. C., Hall, D. T., Krisko, P. H., & Opiela, J. N. (2004). LEGEND—A three-dimensional LEO-to-GEO debris evolutionary model. *Advances in Space Research, 34*(5), 981–986.

Malhotra, A., Gosain, S., & El Sawy, O. A. (2005). Absorptive capacity configurations in supply chains: Gearing for partner-enabled market knowledge creation. *MIS Quarterly, 29*(1), 145–187. https://doi.org/10.2307/25148668

Marwick, A. E. (2013). *Status update: Celebrity, publicity, and branding in the social media age*. Yale University Press.

McDonald, R., & Siegel, D. (1986). The value of waiting to invest. *The Quarterly Journal of Economics, 101*(4), 707–727. https://doi.org/10.2307/1884175

Morris, D. Z. (2016). Leaderless, blockchain-based venture capital fund raises $100 million, and counting. *Fortune*. Retrieved 18 Nov 2025 from http://fortune.com/2016/05/15/leaderless-blockchain-vc-fund/

Myers, S. C. (1977). Determinants of corporate borrowing. *Journal of Financial Economics, 5*, 147–175.

NASA, Bryce Space and Technology. (2024). *SmallSat by the numbers, 2023*. Retrieved 18 Nov 2025 from https://brycetech.com/reports/reportdocuments/Bryce_Smallsats_2023.pdf

Newman, C. J., & Williamson, M. (2018). Space sustainability: Reframing the debate. *Space Policy, 46*, 30–37. https://doi.org/10.1016/j.spacepol.2018.03.001

Paddock, J. L., Siegel, D. R., & Smith, J. L. (1988). Option valuation of claims on real assets: The case of offshore petroleum leases. *The Quarterly Journal of Economics, 103*(3), 479–508. https://doi.org/10.2307/1885541

Pop, V. (2001). The men who sold the Moon: Science fiction or legal nonsense? *Space Policy, 17*(3), 195–203. https://doi.org/10.1016/S0265-9646(01)00023-6

Reuters. (2024, March 5). *Russia says it is considering putting a nuclear power plant on the Moon with China*. Reuters. Retrieved 18 Nov 2025 from https://www.reuters.com/world/russia-says-it-is-considering-putting-nuclear-power-plant-moon-with-china-2024-03-05/

Ryder, S. S., & Hall, P. M. (2017). This land is your land, maybe: A historical institutionalist analysis for contextualising split estate conflicts in US unconventional oil and gas development. *Land Use Policy, 63*, 149–159. https://doi.org/10.1016/j.landusepol.2017.01.026

Saletta, M. S., & Orrman-Rossiter, K. (2018). Can space mining benefit all of humanity? The resource fund and citizen's dividend model of Alaska, the 'last frontier.' *Space Policy, 43*, 1–6. https://doi.org/10.1016/j.spacepol.2018.02.002

Shaw, L. E. (2013). Asteroids, the new western frontier: Applying principles of the general mining law of 1872 to incentivise asteroid mining. *Journal of Air Law and Commerce, 78*, 121. Retrieved 18 Nov 2025 from https://scholar.smu.edu/jalc/vol78/iss1/2

Shull, S. A., et al. (2006). *An integrated modeling tool for sustainable space exploration*. In 57th International Astronautical Congress.

Tourinho, O. A. F. (1979). *The option value of reserves of natural resources* (Research Program in Finance Working Papers, 94). University of California at Berkeley.

Trigeorgis, L. (1995). *Real options in capital investment: Models, strategies, and applications*. Bloomsbury. https://doi.org/10.5040/9780313020827

Tsing, A. L. (2005). *Friction: An ethnography of global connection*. Princeton University Press.

Turkle, S. (1995). *Life on the screen: Identity in the age of the Internet*. MIT Press.

Watts, M. (2004). Resource curse? Governmentality, oil, and power in the Niger Delta, Nigeria. *Geopolitics, 9*(1), 50–80. https://doi.org/10.1080/14650040412331307832

Wilkinson, L. (2015). *Charting a course for effective scientific communication: Balancing accuracy and promotion around the Virgin Galactic crash*. In IEEE International Professional Communication Conference (IPCC).

Wrench, J. G. (2019). Non-appropriation, no problem: The Outer Space Treaty is ready for asteroid mining. *Case Western Reserve Journal of International Law, 51*, 437.

Third World Approaches to Space Mining, Equity, and Erosion of the Common Heritage of Humankind

Adithya Variath and Farhanaz Hazari

Abstract Space will play a critical role in mitigating challenges ranging from climate, and defence, to supply chains. The development of 'space for earth' and 'space for space' technologies that deal with resource utilisation from the Moon and other celestial bodies has triggered multi-billion-dollar investments both from the public and private sectors. Articles I and II of the 1967 Outer Space Treaty (OST) read that "the exploration and use of outer space shall be carried out for the benefit and in the interests of all countries" and it is "not subject to national appropriation." However, these principles are at loggerheads due to the rapid development of space-mining activities. Several space-faring states have introduced national regulations legitimising commercial appropriation of space resources. While the delegation of appropriation rights to commercial/private actors triggers a larger debate around property rights, i.e., can a nation grant a right of appropriation over global commons? Commercial actors argue that international space law does not prohibit extraction explicitly, and space resources are not outer space, the Moon, and other celestial bodies per se. In this context, it is critical to review whether international space law, which entrenches space as a 'province' and 'common heritage' of humankind, provides a mechanism to ensure equitable distribution of benefits accruing from space resources. This chapter locates outer space as an inaccessible domain for the third world and critically reviews how the concentration of technologies amongst a few hegemonic actors makes the 'global commons' inequitable and inaccessible. The chapter draws upon the Third World Approaches to International Law (TWAIL) literature to test whether the principle of common heritage survives in the current era of commercial space mining.

Keywords TWAIL · Global commons · Resource mining · Outer Space Treaty · Common heritage of humankind

---

A. Variath (✉)
Jindal Global Law School, Jindal Global University, Sonipat, India
e-mail: adithya.variath@jgu.edu.in

F. Hazari
Maharashtra National Law University Mumbai, Mumbai, India

H. Jahankhani et al. (eds.), *Space Mining*, Space Law and Policy,
https://doi.org/10.1007/978-981-95-5543-7_2

1 Introduction: Setting and (Un)Settling the Context

In 2023, a team of scientists revisited the Clarion-Clipperton Zone (CCZ), a vast stretch of the Pacific Ocean seabed between Hawai'i and Mexico, to study a test mining site initially disturbed in 1979. This expedition offered a rare window into the long-term environmental impacts of seabed mining, allowing researchers to assess the ecological scars left by human intervention over a span of 44 years. Their findings, published in *Nature*, revealed that reveal that "areas affected by plumes from this small-scale test have limited detectable residual sedimentation impacts with some biological assemblages similar in abundance compared to control areas after 44 years. Although some aspects of the modern collector design may cause reduced physical impact compared to this test mining experiment, our results show that mining impacts in the abyssal ocean will be persistent over at least decadal timeframes and communities will remain altered in directly disturbed areas, despite some recolonisation. The long-term effects seen in our study provide critical data for effective management of mining activities, if they occur, including minimising direct impacts and setting aside an effective network of protected areas" (Jones et al., 2025). The report shows that propulsion-based mining equipment used at the time had caused significant and lasting changes to the seabed sediments, even though it differed from the collector systems being proposed today (Jamasmie, 2025). More alarmingly, the research shows that in some instances profound effects on deep-sea fauna, demonstrating the slow pace of biological recovery in such fragile ecosystems (Cremers, 2023; Jaeckel, 2017; Kim, 2017; Matsumoto et al., 1984).

Under Article 136 of the United Nations Convention on the Law of the Sea (hereafter 'UNCLOS'), the Area and its mineral resources are recognised as the common heritage of humankind (United Nations Convention on the Law of the Sea [UNCLOS], 1982). All activities related to the exploration and exploitation of minerals in the Area are governed by the International Seabed Authority (ISA), which is responsible for implementing necessary measures to safeguard the marine environment from any detrimental impacts resulting from mining operations (Auburn, 1971; Henkin, 1969; Sun, 2023a, 2023b). Scholars argue that "The decision whether, and under what conditions, to mine the deep seafloor does not occur in a vacuum. While the governance of our ocean commons is notoriously fragmented, most of the 168 member states of the ISA are bound by a raft of international legal obligations beyond UNCLOS. These obligations are binding in their own right. Their importance is indeed recognised in UNCLOS Article 237 in relation to other environmental law regimes, and these obligations are relevant to any interpretation of UNCLOS" (Singh & Jaeckel, 2024). However, States and commercial actors have shown further urgency on deep-sea mining, highlighting that our quest for resources under the banner of sustainability or energy transition is creating new frontiers of environmental risk. Despite clear warnings, the pursuit of alternative resource frontiers continues unabated. It is a demand driven by greed, politics and technology. While it has been written by many on the story of the sea, humanity has found its next target, the sky. The next age of resource extraction will be the story of mining the sky.

The extraction of resources from the Moon, asteroids, and other celestial bodies, once the realm of speculative fiction or scientific amusement, is rapidly transforming into a domain of commercial and strategic investment (Foust, 2025). This is not just a result of curiosity or exploration but of economic, industrial, and energy imperatives. This has led to space being reimagined as a "New Scramble", with studies saying the potential of rare earth elements to water ice and from platinum group metals to the elusive helium-3 isotope. Harrison Schmitt, the last astronaut to walk on the Moon in 1972 and a trained geologist, argued as early as the 1980s that lunar resources, particularly helium-3, could revolutionise terrestrial energy systems. In his book, *Return to the Moon*, he proposed a blueprint for tapping the Moon's potential, especially as a source of clean fusion energy (Schmitt, 2007). A 1988 NASA report on helium-3 highlighted its potential for use in nuclear fusion reactors (NASA, 1988). One of the present authors has argued elsewhere, "Helium-3 has the potential to become one of the most important sources of clean energy. As a non-radioactive isotope, it could serve as an ideal fuel for fusion reactors" (Variath, 2024).These ideas are in their initial experimental stages; their success would make it a viable source of energy in the near future. Private companies have and are investing millions of dollars into these energy transition aspirations.

Helium-3 has been viewed by some scientists as a viable non-radioactive fuel for future nuclear fusion reactors. As fossil fuels become politically and environmentally untenable, and first-generation renewables struggle with scalability, helium-3 has re-emerged as a futuristic, if speculative, solution to global energy challenges. This has catalysed a new wave of investment from both states and private actors. Companies like *Interlune*, led by ex-Blue Origin engineers, are exploring ways to commercialise helium-3 extraction (Choucair, 2025). Others also target water ice as a critical resource for sustaining off-world missions and manufacturing propellants in space (Byers & Boley, 2023). Apart from the Moon, some asteroids have attracted considerable interest due to their high concentrations of rare metals. For example, the metallic asteroid *16 Psyche*, estimated to be 200 km wide and composed of iron and nickel, may hold resource values worth trillions of dollars, potentially exceeding the global GDP (Sohn, 2023). Other celestial bodies offer deposits of platinum, iridium, osmium, and palladium, which are increasingly rare and expensive on Earth but crucial for electronics, medical devices, and clean energy technologies.

The primary concern is the economic logic of such ventures, especially given astronomical costs, technological uncertainties, and legal ambiguities. However, major space exploration powers have been drawn into the capitalist logic of extraction and accumulation. Apart from missions like NASA's Artemis programme, India's *Chandrayaan-3* and China's *Chang'e* series have reflected strategic foresight into the exploitation of outer space (Kanu et al., 2024; Wang & Liu, 2016). Data from the Lunar Reconnaissance Orbiter and India's *Chandrayaan-1* offers the possibility of supporting sustainable human presence and producing rocket fuel. Meanwhile, Europe, Japan, and emerging space powers such as the UAE and South Korea have announced initiatives to develop lunar missions that align with their economic and geopolitical interests in space. While these may sound amusing, there lies a more complex and contentious reality of the lunar economy.

Private companies have risen and fallen in the race to monetise space. The technological and environmental challenges remain challenging. Technical difficulties range from lunar dust being highly abrasive, temperatures swinging from searing heat to cold, and radiation levels being extreme. Mining in such alien environments requires unprecedented innovations in robotics, material science, and engineering. While these technical difficulties remain a debate for another chapter, this chapter delves into more profound legal, ethical, and geopolitical questions. This chapter attempts to answer some difficult questions that remain to be answered in this trillion-dollar future of space mining: Who owns these resources? Who benefits from them? And who bears the costs? These are not just academic queries but questions of global justice and historical memory. This chapter attempts to answer these questions through the lens of power, equity, and resistance.

These questions are not new. They echo earlier concerns raised during the formulation of the UNCLOS, particularly in relation to the deep seabed. In 1967, Maltese Ambassador Arvid Pardo gave a historic address at the United Nations, proposing that the deep seabed be declared the "common heritage of mankind" (Dingwall, 2021a, 2021b; Joyner, 1986; Mickelson, 2019; Sun, 2023a, 2023b). His vision was grounded in the fear that powerful industrial states would appropriate the seabed's wealth for themselves, leaving developing nations excluded from the benefits of technological progress. Ranganathan writes: "Pardo's speech (and follow-up activities) introduced a new legal concept, the common heritage principle; it contributed an issue that became important to the NIEO agenda; and it catalysed the demand for renegotiating the law of the sea as a whole" (Christol, 1999; Ranganathan, 2021). Pardo's intervention called for a new international legal order, one that would democratise access to resources, redistribute wealth, and protect the marine environment.

Today, a similar crossroads confronts us in outer space. The Moon Agreement, though poorly ratified, proposed that the Moon and its resources should be the "common heritage of mankind," unlike the Outer Space Treaty (hereafter 'OST'). The resurgence of nationalist space policies, coupled with bilateral agreements like the Artemis Accords, has raised concerns about a return to exclusionary practices. For scholars of the Third World Approaches to International Law (TWAIL), this moment is deeply significant. The same structural asymmetries that plagued Earth-bound resource regimes, colonialism, economic dependency, and epistemic marginalisation risk being replicated in space. Outer space today is a contested legal and political terrain, where the Global South must once again fight for inclusion, recognition, and equity. This chapter engages with these dynamics, examining the political economy of space resource mining through a TWAIL lens. It interrogates the legal frameworks that govern outer space, critiques emerging governance regimes from a postcolonial standpoint, and explores alternative visions of equity and sustainability in space.

2 Answering Why(s) and How(s) Behind the Exploitation of Space Resources

The concept of harnessing space-based resources to support human presence beyond Earth, whether for long-term exploration, settlement, or commercial gain, has been explored for decades. In fact, efforts to extract oxygen from lunar soil were underway even before the first Moon landing. This practice, known as *In Situ Resource Utilization* (ISRU), involves the extraction and use of both natural and discarded materials found in space to produce valuable resources for further application (Jakhu et al., 2016). In 2012, two US-based companies: *Planetary Resources, Inc.* and *Deep Space Industries, Inc.* announced their intention of mining asteroids. While it sparked some debate, it was clear that these were in line with the intention of the US Government. In 2015, the US passed its legislation legitimising this vision, Title IV of the 2015 Commercial Space Launch Competitiveness Act (CSLCA) giving the right of US private entities to engage in the recovery, appropriation, and use of space resources for the purpose of profit (Commercial Space Launch Competitiveness Act, 2015). On April 6, 2020, the White House issued an Executive Order on Encouraging International Support for the Recovery and Use of Space Resources (White House, 2020). Tronchetti and Liu write that "Executive Order does not merely repeat the provisions of the 2015, Act but it goes well beyond them by (a) addressing the status of outer space, (b) explaining the US position toward the Moon Agreement, and (c) laying down the US approach toward international cooperation. Most importantly, the Executive Order is relevant when considered in connection with the fundamental question of how new rules governing the recovery and use of space resources should be developed" (Tronchetti & Liu, 2021). While the legal regimes will be discussed in the next part of this chapter, these developments showed the role of the Government in supporting these initiatives and finding value in them. Since then there have been many startups across the world with similar ISRU plans to exploit water, solar wind-derived volatiles (like hydrogen, helium, carbon, and nitrogen), metallic and mineral-rich regolith, atmospheric elements, continuous solar energy from space, as well as human-generated waste and defunct equipment.

As per a document published by NASA, "ISRU encompasses a broad spectrum of scientific and technical fields, with five key domains critical to human space missions and the commercialisation of space: 1. Resource Characterization and Mapping, 2. In Situ Consumables Production, 3. Civil Engineering and Construction, 4. In Situ Energy Production and Storage, and 5. In Situ Manufacturing" (Sanders, 2015). In its broad sense, these practices mirror many aspects of Earth-based industries, especially in sectors such as food and water systems, mining, infrastructure development, and energy. Among the most frequently cited examples of potential targets for resource exploitation is the asteroid *16 Psyche*, estimated to span approximately 200 km in diameter and potentially composed of up to 50% metal (Schilling, 2023). If accurate, this composition suggests it could contain amounts of iron and nickel equivalent to many times the annual global production, representing an unprecedented reserve of industrial metals. But metallic wealth is not limited to iron and nickel alone.

Certain asteroids are said to contain rare and precious elements scarcely found on Earth, such as platinum, iridium, osmium, and palladium. These elements are indispensable in a wide array of high-tech applications, including catalytic converters, advanced medical devices like pacemakers and implants, and, most importantly, electronic components essential to modern computing and communications infrastructure. Closer to Earth, the Moon is increasingly at the centre of strategic interest, not solely due to its proximity, but because of its abundance of two critical resources: water and helium-3. However, these highly ambitious programmes come with their own scientific and technical problems. Scholars writing about the complexity of these programmes argue that: "the lack of convection in microgravity, problems in mixing and the different behaviours of liquids and solids at liquid–solid interfaces all combine to create complex problems in optimizing mining operations on or near asteroids. In addition to mining, understanding the interactions between minerals, microbes and liquids in space has applications to other diverse problems including life-support systems with a microbiology component that use local regolith as feedstuff, and the behaviour of liquids and organisms in filter systems in space" (Raafat et al., 2013).

In light of such scientific questions, various countries have intensified their space exploration agendas by investing more resources into scientific research. For example, the Artemis programme, spearheaded by NASA, marks the United States' renewed commitment to lunar exploration, seeking not only to return humans to the Moon for the first time since 1972 but also to lay the groundwork for Mars missions. As per recent reports, the US Department of Defence budget request includes 33.7 billion dollars for space programmes in fiscal year 2025 (Clark, 2024).

Despite surging interest and innovation, space resource extraction remains fraught with volatility and financial uncertainty. Many pioneering private companies have either failed to secure sufficient funding or dissolved after failing to raise adequate capital. Despite this, the private sector remains undeterred. Japanese firm *iSpace* aims to provide commercial access to lunar resources, particularly water and minerals (iSpace, 2024). California-based *Offworld* is developing multi-purpose industrial robots capable of mining not just Earth, but also asteroids, the Moon, and Mars (Offworld, 2023). In the UK, the Asteroid Mining Corporation is working on its "El Dorado" satellite intended to perform a spectral survey of over 5,000 asteroids to identify those with high mining potential.

The acceleration of "NewSpace" activities driven by falling launch costs, reusable rockets, and mass-produced satellite components is reshaping the space economy. OECD defines the Space Economy as "the full range of activities and the use of resources that create value and benefits to human beings in the course of exploring, researching, understanding, managing, and utilising space" (Organisation for Economic Co-operation and Development [OECD], 2012). Once dominated by state agencies, the space economy now features an ecosystem of private enterprises engaged in orbital tourism, manufacturing, satellite services, and resource exploration. Current projections suggest that this rate of growth could mean that the space economy will reach $1 trillion by 2040 (McKinsey & Company, 2024).

Despite the immense economic implications, space mining currently operates within a loosely defined legal and regulatory framework. The absence of comprehensive global governance increases the risk of friction. OST, Tronchetti writes, "represents the most important legal instrument of the system of space law establishes principles applicable to all activities to be carried out in the space environment, does not contain any specific reference to the use of space resources. Not even the term "exploitation" is mentioned in the Treaty. This does not mean, however, that the Treaty's principles are not applicable directly or indirectly to the exploitation of extraterrestrial resources" (Tronchetti, 2009). The treaty has often seen deviations not only among spacefaring nations but also between state and non-state actors, particularly in an era marked by renewed geopolitical competition. There is no doubt that establishing new norms could reduce the likelihood of disputes, foster collaboration, and promote the responsible and efficient exploitation of extraterrestrial resources.

Drafted during the early days of space exploration, the OST lacks the specificity necessary to address the complex realities of contemporary space endeavours. Article I of the OST states that: "The exploration and use of outer space, including the moon and other celestial bodies, shall be carried out for the benefit and in the interests of all countries, irrespective of their degree of economic or scientific development, and shall be the province of all mankind" and Article II states that "Outer space, including the moon and other celestial bodies, is not subject to national appropriation by claim of sovereignty, by means of use or occupation, or by any other means." The 1979 Moon Agreement attempted to fill some of these gaps by declaring lunar resources to be the common heritage of humankind, not subject to appropriation. Article 11(5) of the Moon Agreement reads that "States Parties to this Agreement hereby undertake to establish an international regime, including appropriate procedures, to govern the exploitation of the natural resources of the moon as such exploitation is about to become feasible" (Agreement Governing the Activities of States on the Moon and Other Celestial Bodies, 1979). However, the treaty has only 17 parties, with none of the major space powers signing it.

NASA is building a new form of consensus through the Artemis Accords, a soft law that outlines a shared commitment to principles rooted in the OST. Section 10 of the Artemis Accords overtly deviates from OST's vision to legitimise space resource utilisation by reinterpreting Article II of the OST to mean that: "The Signatories emphasize that the extraction and utilization of space resources, including any recovery from the surface or subsurface of the Moon, Mars, comets, or asteroids, should be executed in a manner that complies with the OST and in support of safe and sustainable space activities" (Treaty on Principles Governing the Activities of States in the Exploration and Use of Outer Space, 1967) [OST]. It further states that "the Signatories affirm that the extraction of space resources does not inherently constitute national appropriation under Article II of the OST and that contracts and other legal instruments relating to space resources should be consistent with that Treaty." As of April 8, 2025, Bangladesh became the 54th nation to sign the accords. In this context, Deplano argues that "Artemis Accords turn out to be the cusp of a wider trend in the law-making process for outer space, sitting at the crossroads of multilateralism and unilateral national legislation" (Deplano, 2021).

At the national level, countries have begun to unilaterally craft laws to legitimise and incentivise space resource extraction. Following the 2015 U.S. Commercial Space Launch Competitiveness Act of 2015, nations like Luxembourg, Japan, and the UAE have introduced domestic legislation supporting space mining ventures. Meanwhile, Russia and China have their own ambitions but have criticised the Artemis Accords on several fronts. China has openly criticised aspects of the Accords, such as the proposed establishment of safety zones, while Russia has dismissed the framework as overly aligned with American interests. The lack of consensus exposes the larger problem of power politics further complicating the issues of governance of space resources. Without cohesive global action, the emergence of overlapping conflicting regulatory regimes appears inevitable, heightening the risk of conflict in space. There is a need to shape a rules-based order for space resources. The next part of this chapter delves into the discussion of the fragmentation of the laws on space resource utilisation.

3 A Critical Review of Space Resource Utilisation and Mining Regulations

No nation has been as deliberate, aggressive, and self-serving in shaping the future of space mining as the United States, which has taken an uncompromising "*first-come, first-served*" approach, discarding multilateralism in favour of capitalist expansionism. At its core, the US represents a radical departure from the original vision of space as the Common Heritage of Humankind (CHH). Instead of embracing a cooperative framework where the benefits of space resources are distributed equitably, the U.S. has systematically worked to dismantle any international restrictions that might prevent American Multinational Corporations from claiming vast amounts of extraterrestrial wealth. It can be called as a new form of economic imperialism in orbit.

In 2015, the US Commercial Space Launch Competitiveness Act (hereafter 'SPACE Act') was the first major signal that the U.S. was abandoning any pretence of collective governance (U.S. Commercial Space Launch Competitiveness Act, 2015). Under this legislation, American companies can mine, extract, and sell extraterrestrial resources with full legal backing from the US government. The message was clear that space resources are up for grabs, and whoever gets there first can own and profit from them. Additionally U.S will protect the corporation's rights to these resources even if there is no global consensus. Silence on any obligation to share benefits reaffirms the 'dominance' of the US imperial approach to space mining. Section 51303 on Asteroid resource and space resource rights states "A United States citizen engaged in commercial recovery of an asteroid resource or a space resource under this chapter shall be entitled to any asteroid resource or space resource obtained, including to possess, own, transport, use, and sell the asteroid resource or space resource obtained in accordance with applicable law, including the international obligations of the United States" (NASA, 2020).

The SPACE Act was nothing short of a legislative usurpation, especially one that directly conflicts with the spirit (if not the letter) of the Outer Space Treaty, 1967, which explicitly bans national appropriation of celestial bodies (OST, 1967, art. II). Nevertheless, the U.S has conveniently reinterpreted the clause to mean that while a 'country' cannot claim the Moon or an asteroid as its own, a 'company' can claim and profit from anything it extracts, thereby beginning the period of off-earth industrialisation. Framed as a set of principles for '*peaceful*' space exploration, the Artemis Accords fundamentally prioritise national interest over holistic global development (NASA, 2020). Other than departing from the prohibition of national appropriation of celestial bodies, the Accord, under the disguise of adhering to the Outer Space Treaty, has established the concept of 'safety zones' in Section 11 around the mining and exploration sites, essentially raising concerns about these zones evolving into *de facto* territorial claims.

The Artemis Accords—Section 10 is where the US engages in diplomatic rebranding, i.e. instead of outrightly challenging Article II of the OST, it subtly shifts its approach without ever admitting to it. By declaring it does not "inherently constitute national appropriation," it introduces a legal playground of just the right amount of vagueness to provide flexibility, but firm enough to set a precedent. Section (2) states "The Signatories emphasize that the extraction and utilization of space resources, including any recovery from the surface or subsurface of the Moon, Mars, comets, or asteroids, should be executed in a manner that complies with the Outer Space Treaty and in support of safe and sustainable space activities. The Signatories affirm that the extraction of space resources does not inherently constitute national appropriation under Article II of the Outer Space Treaty, and that contracts and other legal instruments relating to space resources should be consistent with that Treaty." It is not just a matter of semantics but a larger case to shape a precedent at a time when space law is still in its infancy and national regulations remain largely unformed.

Without an internationally recognised governing framework, disputes arising out of the overlapping of 'safety zones' by other nations, the Accord fails to address a clear adjudicating mechanism in cases of potential military confrontations over space resources. While the Accord does not *prima facie* violate OST, it, however, encourages more countries to similarly create parallel legal frameworks suitable for national objectives. This also advances bilateral agreements that sidestep the principles of the 1979 Moon Agreement, which seeks to ensure that the benefits derived from space resource extraction are shared equitably among all nations, particularly those without independent space programmes.

The essence of the Moon Agreement is to prevent the monopolisation of space resources by a few powerful nations or private corporations and to ensure that space remains a global common to mankind. However, the US, having never been a signatory to the Moon Agreement, rejects this principle and instead promotes a national policy favouring commercial space activities. What the US is doing here is not diplomacy but coalition-building for a corporate-led space economy. Rather than fostering a universal, multilateral approach through the Outer Space Treaty, where developing nations could have a meaningful voice, the US has deliberately bypassed these global discussions. Instead, it has crafted a legal framework on its own terms, aligning only with select allies through the Artemis Accords and unilateral executive policies. NASA, not being a law-making body, framed the Artemis Accords as a political commitment rather than a legally binding treaty, which can also be viewed as a strategic move to avoid direct violations of international space law. However, with a growing number of signatories, the Accord is evolving into an "instant custom," a backdoor method of reshaping space law without formal international consensus. This approach subtly erodes the foundational principles of the Outer Space Treaty while rendering the Moon Agreement increasingly irrelevant. Far from mere co-operation, the widespread adoption of the accord signals a calculated shift towards a US-centric framework, setting a precedent where national interests dictate space governance, rather than established multilateral treaties.

Following the US legislation, the Luxembourg Government first sowed the seeds of space mining in early 2016 by announcing a programme aimed at bringing back minerals from space. Being the first European nation to secure property rights for space resources, Luxembourg has been a major force in space exploration and mining extractions (Luxembourg Law on the Exploration and Use of Space Resources, 2017). Having two key legal frameworks for space activities, Luxembourg is set to become a hub for commercial space mining.

The 2017 Law on exploration and use of space resources established that space resources are capable of being owned (Art. 1); however, authorisation is a necessary step and begins to lay down a structure for the same (Art. 2). The law prioritises commercial incentives, fostering investment in space mining, and it lacks clear safeguards for environmental concerns, liability enforcement, and global equity. The 2020 Law on Space Activities establishes an investment-driven approach towards structuring authorisation and supervision of space activities conducted by private operators (Loi du 15 décembre 2017 sur les activités spatiales, 2017). As an incentive, it also provides tax exemptions related to insurance contracts for registered space objects, thereby easing financial constraints because even in space, risk management comes at a price (Luxinnovation, 2017). Building upon these developments, it also established the European Space Resources Innovation Centre (ESRIC) in November 2020 to further scientific, technical, and economic research on space resources. ESRIC aims to become a globally recognised centre of expertise, supporting both human and robotic exploration missions and fostering an in-space economy. To drive innovation, ESRIC recently launched the world's first start-up accelerator dedicated to space resources. This programme helps early-stage ventures validate their technology and develop sustainable business models for both terrestrial and space applications (Japan Act on Launching of Spacecraft, 2016). The creation of ESRIC, while framed as a collaborative effort, may ultimately reinforce Western dominance in space governance by prioritising the interests of Luxembourg, the European Union, and Western spacefaring nations, potentially marginalising those nations that lack the infrastructure or investment capacity to compete. Rather than advocating for a multilateral, redistributive approach that could benefit non-spacefaring nations, proponents of Luxembourg's approach may argue that ESRIC merely promotes international collaboration and that private investment in resource utilisation is already permissible under space law. From a TWAIL perspective, the absence of strong multilateral agreements or binding redistributive measures ensures that only a small group of technologically advanced nations will dominate space mining. Without inclusive governance frameworks, space may become another site of resource exploitation by wealthier states, replicating patterns of economic inequality seen in past colonial and postcolonial extractive industries.

Japan's Basic Space Act of 2008 has also built a robust legal framework for space exploration. It established the guiding principles of Japan's space policy, outlining government responsibilities in fostering space development. What kick-started as a regulation for 'improving the lives of citizens' as under Article 3 of the Basic Space Act, took a turning point and shifted its approach towards commercial viability and resource exploitation as evidenced in the subsequent enactment of the Act on Launching of Spacecraft and Control of Spacecraft in 2016 and the Act on the Promotion of Business Activities for the Exploration and Development of Space Resources in 2021 (Japan Act on the Promotion of Business Activities for the Exploration and Development of Space Resources, 2021). The 2016 Act imposed a stringent licensing requirement, compelling any individual seeking control over a Japan-registered spacecraft to secure approval from the Prime Minister. The 2021 Act further entrenched this regulatory grip by making such a licence a prerequisite for acquiring ownership of mined space resources. Far from fostering an open and co-operative space economy, this regulatory structure entrenches a model of resource extraction that prioritises national and corporate dominance over equitable international participation. By consolidating decision-making power within the state, Japan's framework reinforces a system where access to space resources remains dictated by national interests, mirroring historical patterns of resource control on Earth rather than promoting a genuinely inclusive or multilateral approach.

Unlike the US, which has explicitly carved out property rights over extracted space resources, India's stance remains unclear owing to strategic ambiguity on the law. India lacks a dedicated legal instrument governing space mining. However, adhering to the OST, which specifically prohibits national appropriation over celestial bodies, India's recent participation in the Artemis Accord creates an inherent conflict with the present international law on space in the community (NASA, 2023). India's decision to sign the Accords in 2023 proposes a policy change towards approving resource extraction. In 2023, India also opened its space industry to commercial mining engagement through its Indian Space Policy, 2023, which has similar provisions on resource extraction like the US (Government of India, Indian Space Research Organisation (ISRO), 2023). Non-governmental entities (NGEs) were granted right under 4(11) of the policy to "engage in the commercial recovery of an asteroid resource or a space resource. Any NGE engaged in such process shall be entitled to possess, own, transport, use, and sell any such asteroid resource or space resource obtained in accordance with applicable law, including the international obligations of India." India has a bill (not a law) that has the ability to stifle private sector growth by imposing extensive licensing requirements and governmental control over space activities, is the Draft Space Activities Bill, 2017. Another conundrum the draft proposes is that in its definition of 'space activity' it also gives light to procurement of objects for said activity, and interestingly under section 25(2) it suggests that in case any object is developed or generated on board in outer space shall be the property of the government (Draft Space Activities Bill, 2017, Sec 2(f) and Section 25(2)).

4 Deconstructing the Common Heritage of Mankind Debate

Deep ideological and legal differences about how humanity should handle extraterrestrial resources are reflected in the argument between the 'Province of Mankind' and the 'Common Heritage of Mankind' (CHM), which goes beyond simple semantics (OST, 1967). These two terms, while seemingly similar, represent vastly different approaches to governance, ownership, and distribution of benefits from space resources (Bourrel et al., 2016). The "Province of Mankind" concept, as embedded in the Outer Space Treaty, establishes that outer space shall be accessible to all nations and that activities conducted there should benefit all humankind. It does not, however, go so far as to call for a formalised system of regulatory supervision or fair resource distribution. It offers an ideal framework, but because it is not enforceable, different governments are free to interpret it whichever best serves their strategic objectives.

Politically, the phrase "province of mankind" is useful since it is both flexible enough to protect Northern trade and geopolitical interests and ambiguous enough to please newly independent postcolonial states that emerged in the 1960s (Koch, 2008). The OST prohibits national appropriation (Article II), but it lacks enforceable mechanisms for benefit-sharing, equitable access, or technological redistribution, elements central to meaningful postcolonial justice (Gabrynowicz, 1992). The term "province" universalises access, but ignores structural inequities that preclude equal participation by Global South states in outer space activities. The so-called "freedom" of use becomes a freedom only for those with capacity, thereby reproducing global asymmetries in the space economy.

In contrast, the Common Heritage of Mankind (CHM), first codified in UNCLOS 1982 and later extended to outer space via the 1979 Moon Agreement, marks a more radical intervention, at least in aspiration (Yu & Ji-Lu, 2011). Establishing an international regime for managing global commons, such as the deep seabed or celestial bodies, is a positive obligation imposed by CHM. Its key features include non-sovereignty and non-exclusive rights, benefit-sharing with all mankind, with special attention to developing nations, and international oversight through organisations like the International Seabed Authority (ISA).

CHM introduces a redistributive ethic to global resource governance. Its application to space, especially through the Moon Agreement, has faced systematic undermining by powerful spacefaring states, most of whom have refused to ratify the agreement. From a TWAIL standpoint, this rejection is not accidental but structural: CHM threatens the ability of Western actors to unilaterally exploit outer space resources. The refusal to accept CHM in space law mirrors historical patterns, colonial resistance to redistributive sovereignty, Northern opposition to the New International Economic Order, and current resistance to climate reparations (Melchin, 2015). Customary international law, which derives from consistent state practise and legal recognition, suggests that CHM could gain broader acceptance over time, especially as space mining becomes a pressing reality (Goonesekere, 2023). The precedents set by UNCLOS and the ISA provide a strong legal foundation for applying CHM principles to extraterrestrial resource extraction (Something for everyone: Why the United States should ratify the Law of the Sea Treaty, 2010). However, the reluctance of dominant spacefaring nations to commit to CHM in space suggests an unwillingness to relinquish economic and technological advantages, mirroring historical patterns of resource exploitation (Merdekawati et al., 2024).

Even if the Moon Agreement gave CHM an embryonic place in space law, state conduct is undermining its legitimacy, which is made worse by the space resource law's customary vacuum. Claiming that this appropriation does not contradict Article II of the OST, all established national laws permitting private ownership of extracted space resources have interpreted appropriation of resources as different from appropriation of land, which is based on legal sophistry (Arıkoğlu, 2024). However, from a TWAIL standpoint, this is a well-known pattern of legal fragmentation to support the powerful resource extraction outside of any communal regulatory framework. These actions foreshadow a neo-imperial enclosure of the alien commons, in which entities from the Global North profit and extract while claiming legal neutrality.

Customary international law, which requires state practice and opinio juris, is inherently shaped by who gets to act and whose actions are noticed. In outer space, only a few actors, mostly Global North, have the technical means to generate relevant "state practice." The result is hegemonic norm-making that masquerades as neutrality. This undermines the potential of CHM to evolve into a binding customary norm and signals a return to classical imperial legal logics; a shared domain becomes the playground of the powerful under a rhetoric of universality. The ISA has had difficulty striking a balance between the conflicting interests of the global South and wealthy, technologically sophisticated countries (Environmental Justice Foundation, 2023). Like the deep-sea bed, private corporations, especially those from developed countries, could potentially assert rights over large areas of space resources. Space mining has the potential to mimic the neo-colonial trends observed in deep-sea mining, where the wealthy reap disproportionate benefits while the general public receives meagre returns (Pickens, 2024).

One major issue with trying to apply UNCLOS principles to space mining is that there is no physical boundary or central authority in space. In a space that is neither owned nor governed by a single country or group of countries, the idea of a "common heritage," as defined by UNCLOS, becomes far more nebulous and difficult.

The idea of adopting the UNCLOS framework and creating an 'International Space Mining Authority' (ISMA), analogous to the ISA, offers a future where humanity's extraction of space resources is guided by principles of equity, sustainability, and shared benefits. If deployed tactically, an ISMA-like structure, even as a normative skeleton, could serve as a soft-power lever for space-developing nations, a diplomatic mechanism to challenge the jurisprudential vacuum currently exploited by technologically dominant actors like the United States, China, and a small cadre of private aerospace conglomerates. Any ISMA-inspired body can be conceptualised from first principles, taking into account space governance problems like data sovereignty, orbital debris liability, algorithmic extraction rights, and AI-led exploration. The future of space governance will be based on new legal frameworks that view outer space as techno-political infrastructure rather than as territory, rather than on comparisons to earlier oceans.

5 TWAIL Perspectives on Global Commons, Equity and Ethics of Space Resource Utilisation

Non-excludability is the defining feature of Global Commons, and in relation to outer space, free access was the first customary rule of space law (van Eijk, 2025). The governance of the commons has always been subject to contentious debates and difficulties in building an equitable model of governance. Ranganathan writes, "Two prominent concepts shape the regulatory discourse: "the tragedy of the commons," which is the idea that common resources are over-exploited in the absence of regulation while common concerns remain unaddressed, and "the common heritage of mankind," which reflects the idea that some resources belong to us all, including our future generations, and we cannot be denied rights or responsibilities in relation to them" (Ranganathan, 2016). International law's framework with regard to global commons governance has also exposed the unjust role of powerful actors in framing the architecture and institutions of international law. This central critique is particularly relevant in the context of outer space governance and resource extraction, where contemporary legal regimes continue to reflect imbalances rooted in colonial history.

This historical legacy continues to influence modern international legal structures, including those related to space governance. The Third World Approaches to International Law (TWAIL) framework offers a critical lens through which to examine and rethink the governance of space resource mining. At its core, TWAIL highlights how international law has historically been shaped by European imperialist interests, resulting in systemic exclusion, oppression, mercantilism and exploitation of Third World States (Anghie, 2000; García-Del Moral, 2016; Gathii et al., 2016; Mutua, 2000; Sunter, 2007). TWAIL scholars argue that international law remains a tool of neocolonialism, enabling powerful Western nations to maintain control over global resources and institutions (Chimni, 2017, 2022; Enyew, 2022; Gathii, 2022). Despite the decolonisation movements of the twentieth century, the international legal order has failed to adequately redress past injustices or create a truly equitable framework for global governance. It raises larger questions of property rights and who controls and shapes the laws over the commons. Anghie, exposing the agenda of international law, writes:

> From a TWAIL perspective, what is astonishing is that the ingenious expansion of property rights through international law, and through the expansion of private property rights, was simultaneous with the dispossession of entire peoples of their lands, their territories, their very persons. The relationship is almost asymptotic, the property rights of European entities expanding as the non-European peoples were deprived of their lands and means of existence. The racialized dynamic is at its most literal in the practice of slavery, which turned black people into slaves. This dynamic of expansion and retraction, of inclusion and exclusion, is a feature – a fundamental characteristic – of international law that is in many respects ongoing as the management of global commons, outer space and other such arenas is conceptualized principally through the lens of property. (Anghie, 2023)

Inspired by this collective assertion of independence, TWAIL emerged as a scholarly and political movement aimed at challenging the dominance of Western narratives and institutions in international law. In the context of space mining, these historical doctrines resurface in contemporary debates about the ownership and use of extraterrestrial resources. The history of the language and ideology used in space discourse points to manifestations of coloniality in the international space law regime (Adar, 2022). Quijano defines coloniality as "one of the specific and constitutive elements of a global model of capitalist power. It is based on the imposition of a racial/ethnic classification of the global population as the cornerstone of that model of power, and it operates on every level, in every arena and dimension (both material and subjective) of everyday social existence and does so on a societal scale" (Quijano, 2000).

The concept of *terra nullius*, for instance, is eerily echoed in current space policies that treat celestial bodies as empty, unowned territories open for exploitation by technologically advanced nations. Many interests of Global South nations were excluded from the original formulation of the OST. The interests of Latin America, Africa, Southeast Asia, and the Middle East remained outside OST's framework; however, the next generation of space law cannot remain ill-equipped to accommodate their perspectives and priorities with regard to space resource utilisation. The current flow of debates around resource utilisation is perpetuating a regime that legitimises a new kind of exclusion, one not based on technological access and economic clout. Mohan argues, "The existing space law framework, shaped by the interests and perspectives of technologically advanced States, does not adequately address the needs and concerns of these emerging actors. The growing influence and involvement of non-state actors, particularly private entities, in space activities further complicate this scenario" (Vazhapully, 2024). In this sense, the structure of space law reflects and sustains asymmetrical power dynamics. While the legal order of space is draped in the language of universality and peaceful cooperation, its underlying architecture serves the strategic and economic interests of a few powerful States. During the drafting of UNCLOS, "Pardo's fanciful account of the deep seabed created a powerful and soon-realisable utopia revolving around the seabed. It proactively anticipated the future appropriation of the seabed and therefore justified Pardo's proposal to make this area the common heritage" (Dingwall, 2021a, 2021b; Zhou & Xie, 2024). The global community is witnessing history repeating; however, this time, there is no law where similar third-world thoughts can find a home.

TWAIL critiques such legal constructs as inherently unjust, arguing that they ignore the broader implications of equity, historical disenfranchisement, and global distributive justice. As space becomes the new frontier for resource exploitation, the authors urge a re-evaluation of international norms governing such activities. The American unilateralism in the form of the Artemis Accords, for the authors, represents such a manifestation of the hegemonic interest masquerading as a global soft law. From a TWAIL perspective, such initiatives risk entrenching existing power hierarchies and excluding less powerful nations from decision-making processes. Commercial Space Launch Competitiveness Act of 2015 further exemplify unilateral approaches that sideline broader international consensus. Similarly, countries like Luxembourg, Japan, and the UAE have enacted their space mining laws. While these national laws may be innovative, they also raise concerns about the fragmentation of international law and the potential for conflict and inequality in the distribution of space resources. TWAIL's approach to space resource mining can also be a necessary corrective to the dominant narratives in resource mining.

6 Conclusion

There is a need to dismantle the existing Eurocentric approach to resource mining and to offer an alternative that is more inclusive, just, and responsive to the needs of the actors who do not possess the technology today to explore outer space. As Ejik and Aganaba argue "Rather than force Global South states to compromise principles or complicate progress, let us re-examine the fundamental values at stake" (van Eijk & Aganaba, 2021). There is a need for the reinterpretation of existing legal doctrines of space law to rectify the idea of approaching space as a resource open for extraction. A TWAIL framework for space mining would involve a reimagined international legal order that ensures equitable participation of all nations. It would demand co-authorship of legal standards. The norms of space resource utilisation cannot be drafted in the geographies of the West. International space law has to serve as a platform for cooperation rather than domination and to support the sustainable and equitable use of global commons like outer space. In the case of space mining, this means developing mechanisms to ensure that profits from extraterrestrial resources are equitably distributed and that technological advancements are shared with developing countries. It also means recognising the moral and ethical dimensions of space exploration, including the preservation of celestial environments and the prevention of intergenerational harm.

References

Adar, P. A. (2022). Space and the future of humanity: A TWAIL critique of international space law. *TWAIL Review, 3*, 92–115.

Agreement Governing the Activities of States on the Moon and Other Celestial Bodies, 1979, 1363 U.N.T.S. 22, 18 I.L.M. 1434 (1979), Article 11.

Anghie, A. (2000). What is TWAIL: Comment. *Proceedings of the ASIL Annual Meeting, 94*, 39–43.

Anghie, A. (2023). Rethinking international law: A TWAIL retrospective. *European Journal of International Law, 34*, 7–112.

Arıkoğlu, E. (2024). *The concept of the common heritage of mankind and the legal status of outer space in international law*. In Proceedings for the First Symposium on Space Economy, Space Law and Space Sciences. https://iupress.istanbul.edu.tr/tr/book/proceedings-for-the-first-symposium-on-space-economy-space-law-and-space-sciences/chapter/the-concept-of-the-common-heritage-of-mankind-and-the-legal-status-of-outer-space-in-international-law

Auburn, F. M. (1971). The international seabed area. *International and Comparative Law Quarterly, 20*, 1–20.

Bourrel, M., et al. (2016). The common heritage of mankind as a means to assess and advance equity in deep sea mining. *Marine Policy, 95*, 311–316.

Byers, M., & Boley, A. (Eds.). (2023). Space mining. In *Who owns outer space? International law, astrophysics, and the sustainable development of space*. Cambridge University Press.

Chimni, B. S. (2017). Towards an integrated Marxist approach to international law (IMAIL). In B. S. Chimni (Ed.), *International law and world order: A critique of contemporary approaches*. Cambridge University Press.

Chimni, B. S. (2022). The international law of jurisdiction: A TWAIL perspective. *Leiden Journal of International Law, 35*, 389–407.

Choucair, C. (2025, January 28). *Interlune plans lunar mining mission to extract helium-3 for quantum computing*. Space Insider. https://spaceinsider.tech/2025/01/28/interlune-plans-lunar-mining-mission-to-extract-helium-3-for-quantum-computing/

Christol, C. Q. (1999). Arvid Pardo. *PS: Political Science & Politics, 32*, 777–778.

Clark, J. (2024, May 21). *Space officials outline key investments needed to ensure U.S. maintains edge*. U.S. Department of Defense. https://www.defense.gov/News/News-Stories/Article/Article/3783109/space-officials-outline-key-investments-needed-to-ensure-us-maintains-edge/

Commercial Space Launch Competitiveness Act of 2015, Pub. L. No. 114–90, 129 Stat. 704 (2015).

Cremers, K. (2023). *Digging deep: Critical questions remain in the rush to regulate seabed mining*. Institute for Sustainable Development and International Relations (IDDRI).

Deplano, R. (2021). The Artemis Accords: Evolution or revolution in international space law? *International and Comparative Law Quarterly, 70*, 799–819.

Dingwall, J. (Ed.). (2021a). *The deep seabed beyond national jurisdiction and the rise of mining activities within it by non-state actors*. Oxford University Press.

Dingwall, J. (Ed.). (2021b). *The role of the common heritage concept in deep seabed mining*. Oxford University Press.

Environmental Justice Foundation. (2023). *Towards the abyss: How the rush to deep-sea mining threatens people and our planet*. https://ejfoundation.org/resources/downloads/towards-the-abyss-ejf-deep-sea-mining-report.pdf

Enyew, E. L. (2022). Sailing with TWAIL: A historical inquiry into Third World perspectives on the law of the sea. *Chinese Journal of International Law, 21*, 1–30.

Foust, J. (2025, March 17). *The new wave of asteroid mining ventures*. The Space Review. https://www.thespacereview.com/article/4955/1

Gabrynowicz, J. I. (1992). *The "province" and the "heritage" of mankind reconsidered: A new beginning*. In Proceedings of the 2nd Conference on Lunar Bases and Space Activities (CDSITC Aerospace, pp. 691–693).

García-Del Moral, P. (2016). Feminicidio: TWAIL in action. *AJIL Unbound, 110*, 261–266.

Gathii, J. T. (2022). The agenda of Third World Approaches to International Law (TWAIL). In J. L. Dunoff & M. A. Pollack (Eds.), *International legal theory: Foundations and frontiers.* Cambridge University Press.
Gathii, J. T., et al. (2016). Introduction to symposium on theorizing TWAIL activism. *AJIL Unbound, 110*, 255–260.
Goonesekere, N. (2023, November). International treaty negotiations and the status of the principle of the common heritage of mankind: The 2023 BBNJ Treaty text. *Harvard International Law Journal Online.* https://journals.law.harvard.edu/ilj/2023/11/international-treaty-negotiations-and-the-status-of-the-principle-of-the-common-heritage-of-mankind-the-2023-bbnj-treaty-text/
Government of India. (2017). *Draft Space Activities Bill, 2017* (§§ 2(f), 25(2)). https://prsindia.org/files/bills_acts/bills_parliament/1970/Draft%20Space%20Activities%20Bill%202017.pdf
Government of India, Indian Space Research Organisation (ISRO). (2023). *Indian Space Policy 2023.* https://www.isro.gov.in/media_isro/pdf/IndianSpacePolicy2023.pdf
Henkin, L. (1969). International law and "the interests": The law of the seabed. *American Journal of International Law, 63*(1), 1–23.
ispace. (2024, December 17). *ispace obtains license to conduct lunar surface operations from Japanese government for second mission.* https://www.ispace-inc.com/news-en/
Jaeckel, A. L. (2017). Deep seabed mining and the marine environment. In *The international seabed authority and the precautionary principle.* Brill.
Jakhu, R. S., et al. (2016). *Space mining and its regulation.* Springer.
Jamasmie, C. (2025, March 27). *Deep-sea mining scars remain after 40 years, but life returning.* Mining. https://www.mining.com/deep-sea-mining-scars-remains-after-40-years-but-life-returning/
Japan. (2016). *Act on launching of spacecraft, etc. and control of spacecraft* (Act No. 76 of 2016). https://www.japaneselawtranslation.go.jp/en/laws/view/4194
Japan. (2021). *Act on the promotion of business activities for the exploration and development of space resources* (Act No. 83 of 2021, Article 5). https://www.japaneselawtranslation.go.jp/en/laws/view/4332/en#je_at4
Jones, D. O. B., et al. (2025). Long-term impact and biological recovery in a deep-sea mining track. *Nature, 642*, 112–118.
Joyner, C. C. (1986). Legal implications of the concept of the common heritage of mankind. *International and Comparative Law Quarterly, 35*, 190–199.
Kanu, N. J., et al. (2024). An insight into India's Moon mission—Chandrayaan-3: The first nation to land on the southernmost polar region of the Moon. *Planetary and Space Science, 242*, Article 105864.
Kim, R. E. (2017). *Should deep seabed mining be allowed?* Elsevier Science.
Koch, J. S. (2008). Institutional framework for the province of all mankind: Lessons from the International Seabed Authority for the governance of commercial space mining. *Astropolitics, 16*(1), 1–27.
Luxembourg. (2017, July 20). *Law on the exploration and use of space resources* (Article 1). https://space-agency.public.lu/en/agency/legal-framework/law_space_resources_english_translation.html
Luxinnovation. (2017). *Luxembourg sets out to conquer space. Financial Times (Partner Content).* https://www.ft.com/partnercontent/luxinnovation/luxembourg-sets-out-to-conquer-space.html
Matsumoto, W. M., et al. (1984). *Potential impact of deep seabed mining on the larvae of tunas and billfishes.* U.S. Department of Commerce, National Oceanic and Atmospheric Administration, National Marine Fisheries Service, Southwest Fisheries Center, Honolulu Laboratory.
McKinsey & Company. (2024). *Space: The $1.8 trillion opportunity for global economic growth.* https://www.mckinsey.com/industries/aerospace-and-defense/our-insights/space-the-1-point-8-trillion-dollar-opportunity-for-global-economic-growth

Melchin, G. (2015). You can't take the sky from me: A Gramscian interpretation of the common heritage of mankind principle in space law. *Dalhousie Journal of Legal Studies, 24*, 141–158. https://digitalcommons.schulichlaw.dal.ca/djls/vol24/iss1/5

Merdekawati, A., et al. (2024). Common heritage of mankind beyond treaty provisions: Customary or general principle? *Indonesian Journal of International Law, 21*(3), Article 5. https://scholarhub.ui.ac.id/ijil/vol21/iss3/5. https://doi.org/10.17304/ijil.vol21.3.2

Mickelson, K. (2019). Common heritage of mankind as a limit to exploitation of the global commons. *European Journal of International Law, 30*, 635–663.

Mutua, M. (2000). What is TWAIL? *Proceedings of the ASIL Annual Meeting, 94*, 31–40.

National Aeronautics and Space Administration (NASA). (1988). *Lunar helium-3 and fusion power*. NASA.

NASA. (2020, October 13). *The Artemis Accords: Principles for cooperation in the civil exploration and use of the Moon, Mars, comets, and asteroids for peaceful purposes*. https://www.nasa.gov/specials/artemis-accords/

NASA. (2023, June 21). *NASA welcomes India as 27th Artemis Accords signatory*. https://www.nasa.gov/news-release/nasa-welcomes-india-as-27th-artemis-accords-signatory/

OffWorld. (2023, January 26). *OffWorld is now taking development contracts for new AI-powered swarm robots*. https://www.offworld.ai/post/offworld-is-now-taking-development-contracts-for-new-ai-powered-swarm-robots

Organisation for Economic Co-operation and Development (OECD). (2012). *OECD handbook on measuring the space economy*. OECD Publishing.

Pickens, C. (2024, January 29). *Deep-sea mining regulations remain far from complete*. Pew Charitable Trusts. https://www.pewtrusts.org/en/research-and-analysis/articles/2024/01/29/deep-sea-mining-regulations-remain-far-from-complete

Quijano, A. (2000). Coloniality of power and social classification. *Journal of World-Systems Research, 6*, 342–386.

Raafat, K., et al. (2013). The physics of mining in space. *Astronomy & Geophysics, 54*, 5.10–5.12.

Ranganathan, S. (2016). Global commons. *European Journal of International Law, 27*, 693–717.

Ranganathan, S. (2021). What if Arvid Pardo had not made his famous speech?: (False) contingency in the making of the law of the sea. In I. Venzke & K. J. Heller (Eds.), *The making of international law*. Oxford University Press.

Sanders, G. (2015). *Space resource utilization: Near-term missions and long-term plans for human exploration*. NASA. https://ntrs.nasa.gov/api/citations/20150003498/downloads/20150003498.pdf

Schilling, G. (2023, October 6). Metal-rich asteroids like 16 Psyche could be mined for their resources. *BBC Sky at Night Magazine*. https://www.skyatnightmagazine.com/space-science/asteroid-mining

Schmitt, H. (2007). *Return to the moon: Exploration, enterprise, and energy in the human settlement of space*. Springer.

Singh, P., & Jaeckel, A. (2024). Undermining by mining? Deep seabed mining in light of international marine environmental law. *AJIL Unbound, 118*, 72–77.

Sohn, R. (2023, October 12). *Metal asteroid Psyche has a ridiculously high "value." But what does that even mean?* Space. https://www.space.com/psyche-metal-asteroid-composition

Something for everyone: Why the United States should ratify the Law of the Sea Treaty. (2010). *Journal of Transnational Law & Policy, 19*, 357–369. https://www.unclosdebate.org/evidence/99/us-already-bound-common-heritage-mankind-principle-through-customary-international-law

Sun, L. (Ed.). (2023a). Common heritage of mankind and protection of the marine environment. In *International environmental obligations and liabilities in deep seabed mining*. Cambridge University Press.

Sun, L. (Ed.). (2023b). International environmental obligations and liabilities in deep seabed mining: Introduction. In *International environmental obligations and liabilities in deep seabed mining*. Cambridge University Press.

Sunter, A. F. (2007). TWAIL as naturalized epistemological inquiry. *Canadian Journal of Law & Jurisprudence, 20*, 475–505.
Treaty on Principles Governing the Activities of States in the Exploration and Use of Outer Space, Including the Moon and Other Celestial Bodies, Jan. 27, 1967, 18 U.S.T. 2410, 610 U.N.T.S. 205, Article II.
Treaty on Principles Governing the Activities of States in the Exploration and Use of Outer Space, Including the Moon and Other Celestial Bodies, opened for signature Jan. 27, 1967, entered into force Oct. 10, 1967, 18 U.S.T. 2410, 610 U.N.T.S. 205, Article I.
Tronchetti, F. (2009). *The exploitation of natural resources of the Moon and other celestial bodies: A proposal for a legal regime*. Brill.
Tronchetti, F., & Liu, H. (2021). The White House executive order on the recovery and use of space resources: Pushing the boundaries of international space law? *Space Policy, 57*, 101448.
U.S. Commercial Space Launch Competitiveness Act, 51 U.S.C. §§ 51301–51303 (2015).
United Nations Convention on the Law of the Sea. (1982). *1833 UNTS 397, 21 ILM 1261*, Article 136.
van Eijk, C. (2025). The exclusive making of space law. *Leiden Journal of International Law, 38*, 84–112.
van Eijk, C., & Aganaba, T. (2021). Inspired by Africa: A new approach to global space governance. *New Space, 9*.
Variath, A. (2024, October 20). Lunar mining and the erosion of the common heritage principle: What's the future of the Moon Agreement?*Cambridge International Law Journal Blog*. https://cilj.co.uk/2024/10/20/lunar-mining-and-the-erosion-of-the-common-heritage-principle-whats-the-future-of-the-moon-agreement/
Vazhapully, K. M. (2024). Third world approaches to international space law. In B. Sandeepa & B. Bhat (Eds.), *International space law in the new space era: Principles and challenges*. Oxford University Press.
Wang, Q., & Liu, J. (2016). A Chang'e-4 mission concept and vision of future Chinese lunar exploration activities. *Acta Astronautica, 127*, 678–683.
White House. (2020, April 6). *Executive order on encouraging international support for the recovery and use of space resources*. https://trumpwhitehouse.archives.gov/presidential-actions/executive-order-encouraging-international-support-recovery-use-space-resources/
Yu, J., & Ji-Lu, W. (2011). The outer continental shelf of coastal states and the common heritage of mankind. *Ocean Development & International Law, 42*, 317–328.
Zhou, H., & Xie, J. (2024). Revisiting heritage in the ocean: Common heritage of [Hu]mankind, maritime heritage and beyond? *International Journal of Heritage Studies, 30*, 687–702.

The Securitisation of Space Mining. Space Resources Acquisition in Between Geopolitics, Supply-Chain Challenges, and Environmental Risks

Stefania Paladini

Abstract The acquisition and industrial exploitation of resources in/from outer space (i.e., 'space mining') is at present more an ambitious plan than an actual thing, and enormous challenges will have to be met before it becomes reality. And yet, the power politics on Earth is already shaping the sector-to-be, with nation-states' sourcing plans that clash in their trajectories and an increasingly hostile narrative. This is leading to a securitisation of the sector that can transfer dangerous dynamics from Earth to space and threaten the development of the sector itself, which should instead focus on addressing feasibility challenges first and foremost. Moreover, outer space is a fragile environment, and the use of advanced technology could be used in an offensive capacity, leading to cyberattacks and disruptions of space operation. This article investigates the ongoing securitisation and its risks, highlighting risks and challenges, from legal to technical to logistical, and the possible solutions to ensure space sustainability is given the centrality it deserves for a peaceful and fruitful development of the sector.

Keywords Securitisation · Advanced digitalisation · Space-mining supply-chain management · Scenario analysis

1 The Edges of the New Space Economy. An Introduction

In 2024, even in a macro-economic context marked by a slight deceleration in global GDP (estimated at an overall 3.1% growth; ESA, 2025), the numbers of the space industry have kept rising at a steady pace, to no-one's surprise. For several years now the growth of the sector has attracted global attention and interest, from both the public and private sector, including countries that had never before ventured into what many call the "space club."

S. Paladini (✉)
Queen Margaret University Edinburgh, Musselburgh, UK
e-mail: SPaladini@qmu.ac.uk

H. Jahankhani et al. (eds.), *Space Mining*, Space Law and Policy,
https://doi.org/10.1007/978-981-95-5543-7_3

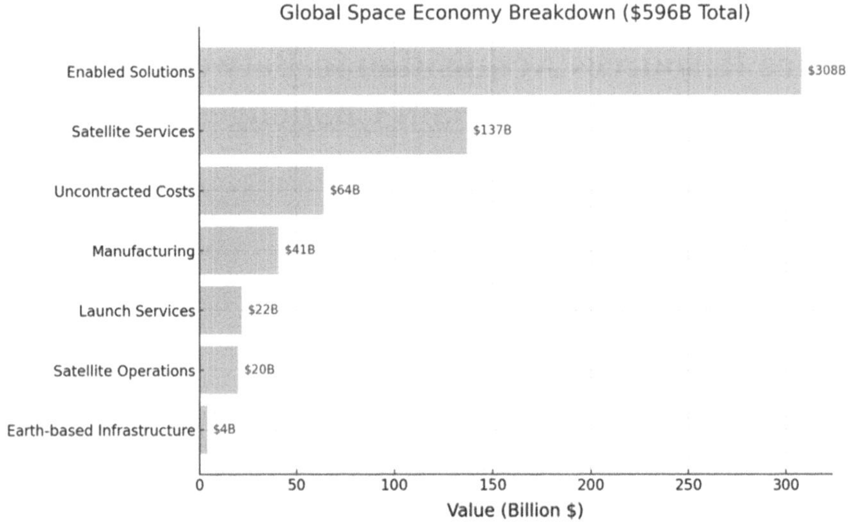

Fig. 1 The value of space economy, 2024 2025 (*Source* Author's elaboration on Novaspace data, 2023–2025)

The 2024 data have returned a clear snapshot of a sector whose growth seems knowing no pauses (Fig. 1) in launches (259), investments and returns, reaching an overall value of 596 billion USD (Novaspace, 2024).

To get a sense of these ascending trends it suffices to look at past numbers.

In 2023, the sector was estimated globally at 570 billion US dollars (The Space Foundation, 2024). In 2009, it amounted to only 200 US billion.

And while there are undoubtedly a lot of complexities in measuring the size and segments of the space sector (OECD, 2014, 2019),[1] its key role in terms of its contribution to the economic growth of leading countries is clear, regardless of the metrics adopted.

This explains why space-active countries are pouring a sizeable portion of their GDP budgets into the sector (see Fig. 2).

In some cases, this percentage is significant by any standard, to reinforce the strategic importance of the space industry in the country's economic policy.

[1] The space sector has always been notoriously difficult to quantify, partly because some segments (e.g., satellites for surveillance and strategic security, launch vehicles for military use) involve data that is often classified and not readily accessible. This is true even for Western countries, while others, like China, has historically treated them as military secrets. Additional complexities arise from non-comparable statistics over time, the inclusion of new market categories as the space industry itself expands, mixed public–private funding, and inconsistent statistical standards between countries. This is reflected in the wide variation in sector estimates depending on the source, which explains the differing figures reported by the Space Foundation, the SIA (Satellite Industry Association), the OECD, the EU/ESA, and by private consultancies, like Novaspace. The ones cited in this report are included in the reference list.

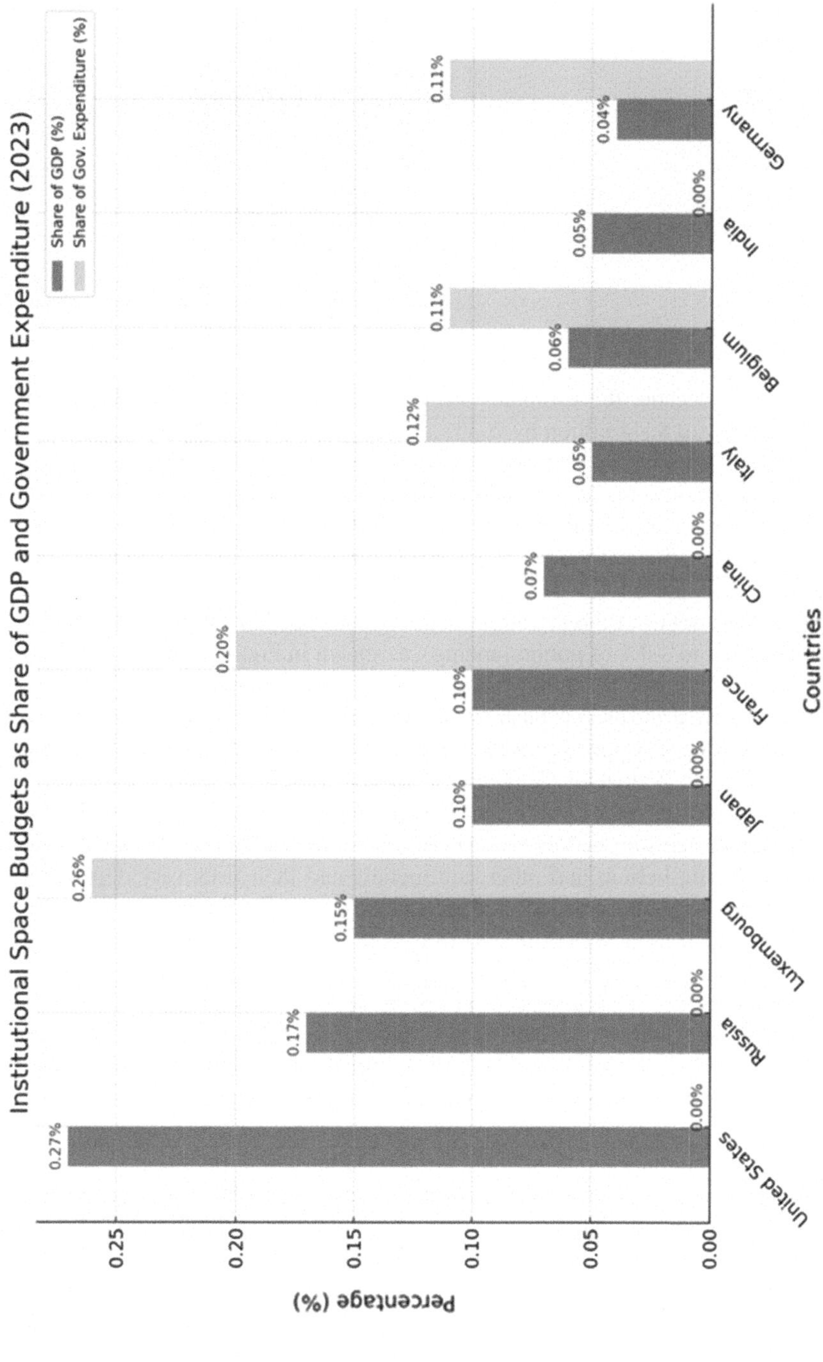

Fig. 2 Institutional space budgets as share of GDP and government expenditures (2023) 2025 (*Source* Author's elaboration on ESA and ESPI data, 2025)

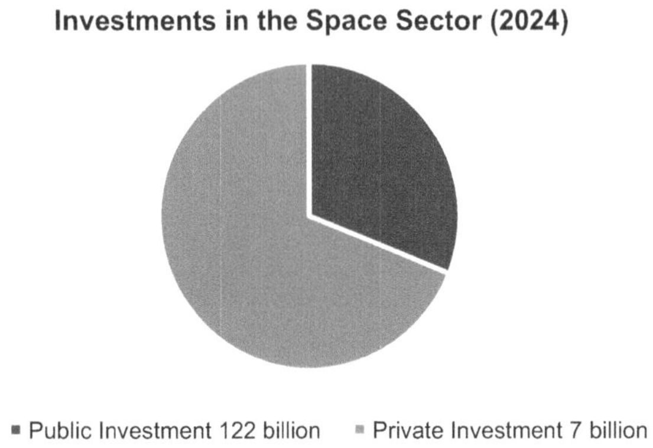

Fig. 3 Investments in the space sector by type 2025 (*Source* Author's elaboration on ESA data,)

For the next decade, the estimates are even more optimistic, with the sector projected to grow to $944 billion by 2033.

And, although competing forecasts do exist (ESA, 2025; ESPI, 2025; Novaspace, 2025), what is certain is that this growth will likely come from an expansion of these existing sectors as much as, and probably particularly, the development of new ones.

The high expectations of the private investors for the future of space industry are illustrated in the chart below.

Notably, the growth of the private investments in 2024 has outpaced the public (+20% compared to +9% of public funding), as shown in Fig. 3.

This is especially true for those segments that right now appear more like a far-away dream than real and present business opportunities, space mining being only one among many (space 3D printing, in-orbit servicing, space tourism, quantum-powered communication, etc (Valuables Consortium, 2021)).

Space mining is based on the idea of extracting resources from asteroids, the Moon, and other celestial bodies. The specific resources/materials (from water to platinum, including helium and other rare metals) and their uses vary depending on the scenarios (use on Earth, on orbiting space stations, on the lunar surface, or even farther), which entail very different evaluations in terms of cost/opportunity and feasibility.

Space mining, it is often argued, could pave the way for alternative sources of resources such as water, rare metals, and minerals, offering possibilities for self-sufficiency in future space missions. As such, it represents one of the most promising frontiers of the space economy, although it is also the one with the most distant time horizons.

While this article focuses specifically on space mining, it is important to say that the segment shares many of the characteristics of the other 'edges' of space economy here mentioned (Paladini, 2019).

This needs to be kept in mind, because what they all have in common is not, or not only, the clear economic gains they promise. It is also their disruptive potential, which will require the adaptation of existing regulatory framework (nothing short of a rewrite in some cases) to meet the economic, environmental, and international security challenges, as well as the integration of new technologies and operational methods.

There are, of course, important differences, the first being the time horizon.

Space tourism, intended here as orbital and suborbital flights for entertainment purposes, is, for instance, more advanced compared to space mining, which remains, at the present stage, more theoretical than anything. This is due entirely to technical reasons, not legal or business ones, although the economic implications of introducing space-derived resources into the terrestrial market could be substantial, potentially triggering financial upheavals.

And yet, nobody doubts space mining is going, eventually, to become a thing.

There is no *if*, only *when*. But the point that matters the most, a point often overlooked, is *how*. Because depending on the way this happens, it will largely affect, or be affected by, the ongoing space industry's power politics.

We have already evidence of an ongoing securitisation, both of space as a whole (although many could argue this is not a novelty. We discuss this claim in the next section) and of space mining in specific.

Based on these premises, this article is structured in the following sections.

After this introduction, Sect. 2 discusses the securitisation of the space sector as part of the more general, and ongoing, geopolitical shift, which sees space as one of the privileged areas of confrontation between (space) powers, with all the risks this entails.

Section 3 focuses more specifically on space mining as one of the segments that have contributed more to this dynamic of securitisation, proposing a conceptual framework for analysis with three main components: technical, economic, and regulatory.

Section 4 offers an overview of what we labelled the two 'feasibility' dimensions, which include both the economic component and the logistical-technical one, while Sect. 5 addresses what is probably the most critical factor of the entire space sector, the regulatory framework and its implications for the wider geopolitical equilibrium.

Section 6 deals instead with the challenges and the risks any serious initiative in space mining entails at environmental level, intended here as a proxy for sustainability of the sector itself. Section 7 discusses what lay ahead for the sector and offers some conclusive observations.

2 An Ongoing Securitisation? Space as a Security Object.

There is no shortage of evidence of the securitisation of the space sector, at least in the terms defined by the Copenhagen School (Buzan & Waever, 2003).[2]

---

[2] The reference here is clearly to the securitisation dynamics, where the 'making' of a security object comes with a 'speech act' by a securitisation actor. This comes with an enhanced attention, generally more resourced devoted to the object but also some unintended consequences, often not in the sense desired by the very securitisation actors. See Buzan and Waever (2003) for more.

And if it is true space used to be at the beginning of the space era (1957) a security object by definition (being an integral part of a military confrontation between USRR and USA), from the 1990s onwards this view held no longer validity with the emersion of the 'commercial space' and the big satellite companies. In the 2000s the sector moved towards an even more integration of the private investors in the different segments of the industry and to an international cooperation among spacefaring nations with few exceptions.

However, things have sensibly changed in the last decade, reflecting a geopolitical shift in global affairs. We only need to look at the various declarations, some of which are from "securitisation actors," to appreciate the extent of this shift.

As early as 2016, the EU Commission declared "*space is becoming a more contested and challenged environment*," (the EU, 2016: 16), in what turned out to be only one among many statements of similar kind.

The literature on space security, still heavily influenced by an IR realist approach (Jakhu & Pelton, 2017; Pfalzgraff, 2015; Schrogl et al., 2015; Sheehan, 2015) supports and explains this interpretation, stating that "geopolitical theory developed for the Earth and its geographical setting can be transferred to outer space" with the "*strategic application of new and emerging technologies within a framework of geographic, topographic, and positional knowledge*" (Dolman, 2002: 23).

Some of the space security aspects identified in the current discourse are military and defence related (such as ASAT, the controversial antisatellite weapons); others are more properly belonging to non-traditional security issues (Deudney & Matthew, 1999; Elliot, 2002; Paladini, 2023), debris *in primis* (more about this in Sect. 6) and non-man-made ones such as space weather.

In this underlying narrative of space security and security in space, space mining is, quite naturally, emerging as one of the most contested areas, if possible, as it comes straight into the—real or perceived—competition for resources that is projected to happen in outer space. If anything, because "*the path of conflict in space and through space starts and ends on the ground*," (Hostbeck, 2015: 56), where clashes on critical materials are at some recent highest levels. And it is not by chance that some of the 'speech acts' that have more contributed to the securitisation of the space sector were actually related to the establishment on Moon outposts on the South Pole[3]: this is where the reserves of water are located, a prime target for space mining.

3 Space Mining: An Analytical Framework

To outline and together analyse space mining as an autonomous, future segment of space industry, we made the choice to use a conceptual model.

[3] See, for instance, the declaration of NASA administrator Bill Nelson, about China's rapid advancement in space capabilities that could provoke a territorial land grab on the lunar surface, much like the tensions that have played out in the South China Sea (Yahoo Finance, 2023).

This is due to two main reasons. The first is their use is well established in methodological terms, counting more than 60 definitions in social sciences (Creswell, 1994; Delcambre et al., 2018; Mylopoulos, 2020; Thalheim, 2018), including in the space sector. They guide "*experimentation by expressing the modelling objectives, and model inputs and outputs*" (Robinson et al., 2015: 12), making them an excellent choice for our purposes.

The second is due to the unique characteristics of space mining, which, even more than other segments, lack hard data on several aspects, including costs, feasibility, and even projected returns, as discussed later on in Sect. 4.

In these conditions, outlining the conceptual aspects seems the only viable way to offer any insight into a sector still largely hypothetical in terms of possible implementations, although not unrealistic.

Figure 4 shows the conceptual model as consisting in macro-dimensions, all considered essential for space mining. Each of these dimensions represents the skeleton of space mining practical implementation. Equally important are their intersections, each of them highlighting a component of relevance to enhance the odds of success. Finally, the core, where all the external layers must be balanced to ensure the sector's long-term viability and, eventually, existence.

More specifically, the model is articulated over three ideal levels, the first involving three dimensions (A), the second, three intersections (B), and a third one centred on a core component (C), which is affected and affects all the others.

The macro-dimensions (A Level) are so defined, and, albeit the model gives them the same visual 'weight', their relevance in practical terms is not the same, as the descriptive below will discuss.

A1. Economic

The economic dimension addresses the financial viability and profitability of space mining ventures. It involves assessing market conditions, the costs of launching and maintaining mining operations in space, the projected return on investment, and the economic/market value of extracted resources. This component evaluates up to which points space mining activities are financially sustainable, attract public & private investments to the sector, and may eventually stimulate an overall industry growth.

A2. Technical—Logistic

Closely related to the Economic is the technical dimension, which refers to the scientific and engineering challenges associated with space mining. This includes the development and deployment of advanced technologies for extraction, transportation, and processing of resources, as well as robotic and automated systems capable of operating reliably in harsh space environments (be they in-orbit manufacturing, in-situ operation on the Moon or asteroid retrieval). Challenges also involve overcoming technical hurdles like low-gravity operations, autonomous robotic control, and reliable energy supplies in remote environments, while controlling for safety.

A3. Legal

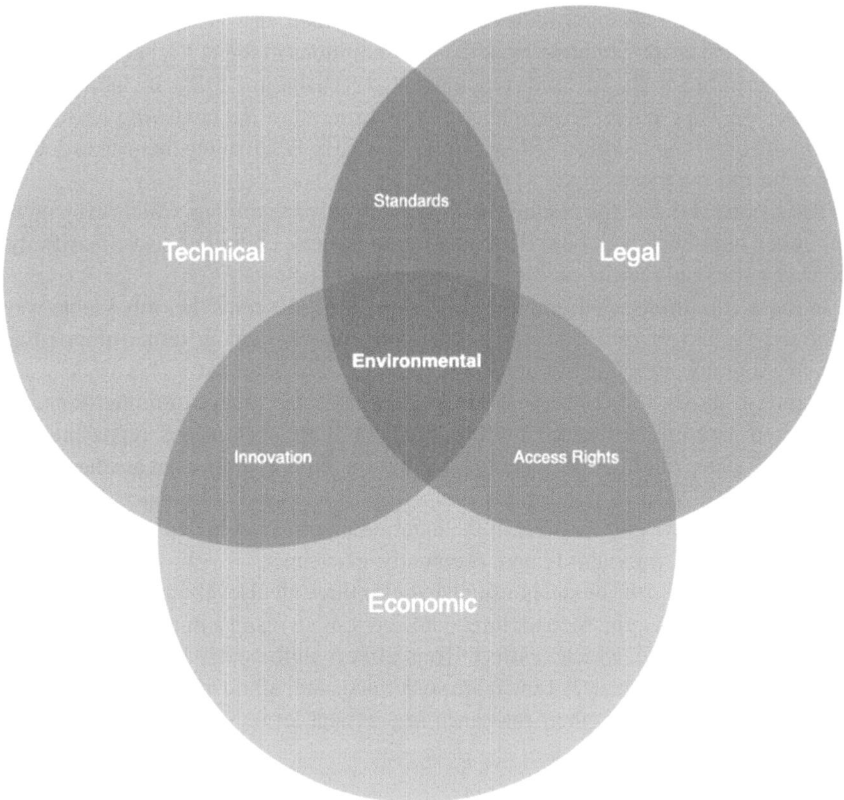

Fig. 4 A space mining conceptual framework (*Source* Author's Elaboration)

On a different level, and more crucial in terms of relevance for the future of the entire sector, and not just space mining, is the regulatory framework. The legal dimension governs the rights and responsibilities related to space mining, including property rights, liability, jurisdictional issues, and international treaties. Establishing clear, globally accepted regulations provides a stable environment for investments and cooperation. This legal clarity is essential to prevent conflicts over resource ownership and ensure compliance with international space laws such as the Outer Space Treaty and the other UN treaties of the 1970s (UNOOSA, 2022).

The B level includes what we can call "the Intersections," and namely:

B1. Technical ∩ Economic = Innovation

Innovation is one positive outcome of the alignment between market incentives and business needs with technological advancements.

The economic drive to maximise efficiency, reduce costs, and improve profitability encourages the adoption of new technologies, regardless of how potentially risky they might sound. Companies and governments invest in R&D to create more efficient extraction techniques, better robotics, and optimised spacecraft design, driven by the economic potential of space mining.

B2. Legal ∩ Economic = Access Rights

This intersection addresses the need for clear property rights and secure access to resources, critical for economic sustainability.

Clearly defined legal access rights provide businesses with certainty, promoting investment and competition. Without legal certainty on the resource destination, the economic incentives weaken, the investment climate is perceived riskier, and the threat of potential disputes over resource ownership (some of which have been already voiced by a few countries; see Sect. 4) could discourage economic development in space mining ventures.

B3. Technical ∩ Legal = Standards

A clear set of industrial standards can and should emerge from the interaction of technical capabilities and legal frameworks. As technical processes for resource extraction and handling develop, standardised procedures and regulations must be implemented to ensure safety, interoperability, and environmental protection, as has happened in other industrial sectors.

Legal mechanisms enforce these standards, ensuring that technical operations adhere to accepted international norms and promoting collaboration between entities operating in space.

C. The C level is The Core: Environment

This specific set is intended here as a proxy for Sustainability, as in the UN COPUOS—*United Nations Committee on the Peaceful Uses of Outer Space—definition: "space sustainability is the ability to conduct space activities indefinitely into the future, ensuring equitable access to the benefits of space exploration and use for peaceful purposes, while preserving the outer space environment for future generations,"(UN COPUOS, sd).*

This component is equally influenced by the economic, technical, and legal dimensions[4] but translating in a far wider perspective, as highlighted in Sect. 6. And while space sustainability rests on the balanced interactions among these three factors, the impacts of wrong choices made at this level are due to affect the entire sector, with crucial externalities to the entire space industry.

---

[4] The importance of the economic dimension is, for instance, evident in other definitions of 'sustainability for the space sector' offered by international actors, such as the space agencies. For instance, *ESA states that "space sustainability refers to the responsible use of space to ensure that future generations can continue to benefit from space-based services and activities. This includes managing space debris, preventing collisions, and minimizing the environmental impact of space missions,"* (ESA, n.d.). *NASA echoes it, although also adding an international cooperation dimension: "Space sustainability involves the preservation of the outer space environment through the mitigation of space debris, the adoption of best practices for space operations, and international collaboration to ensure the long-term viability of space activities,"* (NASA, sd).

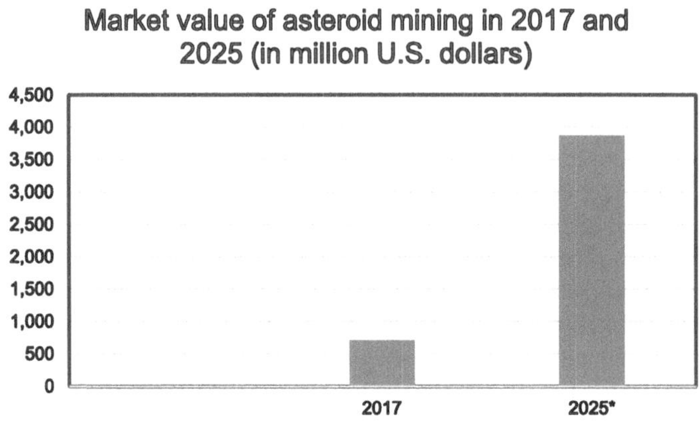

Fig. 5 Market value of asteroid mining (selected years) (*Source* Author's elaboration on Allied Market Research and Statista data, 2025)

The following sections will cover the three dimensions and their intersections as discussed here, to then move to discuss the core dimension of sustainability.

4 The Feasibility Dimensions. The Geo(Economics) of Space Mining and Its Technical and Logistical Challenges.

Any serious analysis of space mining should discuss its feasibility in the first place, which passes, as seen above, through an economic aspect and a technical one. While closely related in many ways, they present a set of risks and challenges of their own.

While the actual market for space mining is still at a conceptual stage, its market value is not, at least if we include in the evaluation the development of prototype, patents, and technologies specifically aimed at enabling space missions for the purpose of gathering raw materials.

An estimate for 2025 is included in Fig. 5, and even more is the projection for the future (not included in the chart), although future projections are more uncertain and diverge substantially depending on the source.

But the corporate market is of certain interest. In addition to the well-known pioneers of space mining, such as Deep Space Industries (later acquired by Bradford Space; SpaceNews, 2019a, 2019b) and Planetary Resources,[5] both of which had strong links with Luxemburg, others have emerged, some of which supported by

[5] Which suffered the same fate of its early competitor, acquired by the blockchain firm ConsenSys in 2018 (SpaceNews, 2019b).

the space agencies—Ispace Europe, Ispace Japan, Lunar Outpost, Masten Space System,[6] and Astroforge[7] among them.

What is currently sparking the biggest controversies, however, is just not the actual value of the resources—although the estimates vary significantly depending on the sources, some of which are even hyperbolic (Space.com, 2023).

It is where to mine them first, and where to put the resource extracted to use.

The first choice to make is between asteroids and the Moon.

Asteroids were originally considered as the first target for mining.

According to Asterank,[8] somehow the starting point of any investigation on space mining, a few possible options are available as a target, depending on what the key objectives of a retrieval mission are. Davida 511 is indicated as the most valuable of them all (with an estimated value of more than 100 trillion US), although the location (in the Asteroid Belt) makes it farther away than others. Psyche makes another target, due to its unique composition in addition to the presence of valuable materials.[9]

Ryugu, already the target of a scientific retrieval mission of Jaxa's Hayabusa 2,[10] is positioned as the most "cost effective," with Bennu, another well-known asteroid in the scientific community (due to NASA's OSIRIS-Rex retrieval mission 2016–2023; NASA, 2023), as a runner-up.

NEO (Near-Earth Orbit) candidates look more accessible in terms of sheer mission planning, although the costs are still consistently high even in this case. Just to give a rough idea of what such an endeavour could entail, the OSIRIS-REx mission cost was about $1.16 billion ($588.5 million on the spacecraft itself, $183.5 million for the Atlas V 411 launch vehicle, and about $283 million over the nine years of mission operations) (The Planetary Society, 2023).

A more viable alternative is, of course, the Moon, starting from the lower mission costs (estimates for the establishment of a lunar landing pad are between $47 million

---

[6] The companies were awarded in 2020 by NASA about 25,000 USD for the collection of lunar samples (SpaceNews, 2020). NASA received 22 proposals overall, 14 of which were rejected either on technical grounds or for cost exceeding the grant limits.

[7] AstroForge raised $40 million more from investors for a mission planned for 2025 at some moment (Wall, 2024), focusing on metals retrieving.

[8] Asterank is an extensive database combining scientific and economic data on more than 600,000 asteroids, aggregating information from reputable sources such as NASA's Jet Propulsion Laboratory (JPL) Small Body Database and the Minor Planet Center. In addition to that, it also offers in-depth details about their physical properties, orbital paths, and projected mining value. Originally developed by Ian Webster, the platform was acquired by Planetary Resources in May 2013, but it remains available up to now for consultation with data over 6,000 asteroids tracked. Their number is however far higher. According to ESA estimates (ESA, 2017) of about 600,000 identified asteroids in our Solar System, more than 16,000 are NEOs. Access the dataset here: https://www.asterank.com/about.

[9] This particular asteroid has been the object of a lot of press attention, with estimates in the order of US$700 quintillion worth of gold and precious metals (Yarlagadda, 2022). In October 2023, NASA sent a mission to Psyche, which is bound to reach the asteroid in 2029 (Warren, 2023).

[10] Japan Aerospace Exploration Agency (JAXA) explored Bennu through its Hayabusa2 spacecraft in between 2014 and 2019, with the sample retrieval successfully reaching Earth by the end of 2020 (NASA, 2025).

and \$229 million, depending on variables like transportation expenses and construction solutions; Metzger & Autry, 2022) to better accessibility and communication modalities.

While it could still provide access to various natural resources (precious ones, like PRG, the platinum group, although to a limited amount compared to the asteroids) and rare minerals, the majority opinion (The Rand Corporation, 2022) is that the most immediately significant resource is likely to be water. Water is incredibly valuable as a source of H and O, that could serve as a rocket fuel.

NASA has long identified ice deposits in deep craters near the Moon's south pole, which is now the target of the exploration efforts of the 2020s exploration missions detailed later in this section.[11]

Another critical decision regards instead where the resources extracted are to be used. While bringing back minerals to the Earth had some obvious advantages, flooding the market with out-of-planet resources would most likely make the prices themselves collapse (Smith, 2019). Moreover, the real gain would be the use of those materials directly in orbit and/or on the lunar soil, as taking payload to orbit is one important mission cost.

The second, important point is more technical and logistical than economic-related, and it concerns the viability of the endeavour at all, which is not, or not only, a matter of costs, but of feasibility of the whole or even part of the complex supply chain that should be put in place to make the retrieval possible.

Some early studies (e.g., Keck Institute for Space Studies, 2012) have offered a mixed picture in terms of actual odds of success. In their simulation, the plan was for an Asteroid Redirect Mission (ARM) capturing a NEO asteroid (approximately 7 m in diameter and weighing 500 tons) to place it in lunar orbit. Among those involved in the study was the CEO of Planetary Resources. The estimated cost for such a mission was \$2.6 billion, based on 2012 U.S. dollars.

Among the preconditions of such missions, there is consensus about three of them: "*Development of a ground-based observation campaign to discover and characterise potential target asteroids for return; development of a sufficiently powerful solar electric propulsion system necessary to transport the asteroid back to lunar orbit in a reasonable time. Establishment of a human presence in lunar orbit to investigate and exploit this resource,*" (Space.com, 2012).

In any case, with the data currently at our disposal, which are rough estimates, it is impossible to say what could be the most suitable option for a start in earnest of retrieval that can be both economically profitable and technically viable. This would depend on a decision previously made on the destination of the retrieved resources, on Earth or in situ.

---

[11] This would in turn become essential even further down the timeline of space exploration. "*This could in turn enable the exploration of more-distant destinations, like mineral-rich objects in the asteroid belt between Mars and Jupiter and potentially lead to vast economic and societal opportunities for the nations involved,*" (The Rand Corporation, 2022).

What is certain is that space players have a few missions in their agenda for the next decade. The Moon figures prominently in them, with the ongoing Chang'e missions (China), which, among others, have as their aims returning lunar samples.

Russia and India have also got their eyes on the lunar soil, with their competing missions Luna 25 (Russia, 2023) and Chandrayaan-3 (India, 2023), both targeting the Moon's south pole. Cislunar orbit, the region of space between Earth and the Moon, represents another crucial strategic area for future space missions, encompassing economic and strategic dimensions, including orbital stations such as the ESA Gateway. Other activities in this area include preparations for establishing lunar surface bases, extraction, and transportation of resources, as well as creating logistical outposts for long-duration missions.

But there is another hurdle still to confront once the prized resources are eventually retrieved. It is the right to legally 'own' and use them.

5 The Regulatory Conundrum

If there is any of the three dimensions that is more closely related in a cause–effect web with the ongoing dialectics of space power competition, this is the regulatory framework (von der Dunk, 2023; Paladini and Castellucci, 2022).

The diagram does not do justice to the prominence this dimension presents over the other two, in terms of possible repercussion on the delicate geopolitical equilibrium in outer space a dispute over resources can entail.

Moreover, it has been, in specific, the perceived need to regulate and establish once and for all what is the legal status of space resources that has provoked the recent round of national space laws (the most relevant of each are listed in Table 1). While not all of them, as we will see, specifically deal with them, the most innovative and groundbreaking do, and will probably offer a strong incentive to all the other players to move in.

It is also important to observe that some of these regulatory efforts are especially important for the future of space mining.

They lay the foundations for what does not contradict Article 2 of the Outer Space Treaty (1968),[12] which prohibits sovereignty claims over celestial bodies, but rather constitutes an important complement to it.

Unsurprisingly, the pioneer in this area was the United States, with its *Space Resource Exploration and Utilization Act* of 2015, which is part of the broader *Commercial Space Launch Competitiveness Act*.[13]

---

[12] Art. 2, OST, 1967: "*Outer space, including the Moon and other celestial bodies, is not subject to national appropriation by claim of sovereignty, by means of use or occupation, or by any other means*".

[13] U.S. Commercial Space Launch Competitiveness Act—Title 51 (Public Law114–90 114th Congress, November 25, 2015) Art. 51 dispone verbatim "*a United States citizen engaged in commercial recovery of an asteroid resource or a space resource under this chapter shall be entitled to any asteroid resource or space resource obtai- ned, including to possess, own, transport, use, and*

Table 1 National space laws (selected countries)

Country	Title of the law	Year
United States	U.S. Commercial Space Launch Competitiveness Act	2015
Luxembourg	Law on the Exploration and Use of Space Resources	2017
United Arab Emirates	UAE Space Law	2019
Japan	Basic Space Law (2008); Space Resources Act	2021
France	Space (Launches and Returns) Act Amendment	2024
United Kingdom	UK Space Industry Act	2018
Italy	Organic Space Law	Senate (2025)
Canada	Canadian Space Act	1990
Portugal	Decreto-Lei n.° 16/2019	2019
New Zealand	Outer Space and High-altitude Activities Act	2017
European Union	***EU Space Law Framework***	***Various***
Multinational	***Artemis Accord***	**2020-**

Source Author's elaboration on law sources (2025)

This legislation essentially states that resources obtained from asteroids or other celestial bodies become the property of those who make the effort to extract them. It allows U.S. public and private actors to use space resources, in what has been called *"the greatest single recognition of property rights in history.*[14]"

Somehow different the case of Luxembourg, the undisputed European leader in space mining legislation, which is at present also the European hub for innovative services in the space sector.[15] As early as 2017, just two years after the approval of U.S. legislation and before the UK's 2018 framework law on the sector, the Grand Duchy passed an innovative law[16] allowing companies registered in its territory to legally extract and use resources from celestial bodies.

The country also established the Luxembourg Space Agency and created a series of investment funds and tax incentives to support companies in the space sector.

While a detailed examination of this legislation is beyond the scope of this article, it is worth noting that Luxembourg's law is more detailed[17] than its US counterpart. It even establishes authorisation mechanisms and, in Article 4, the requirements for

sell the asteroid resource or space resource obtained in accordance with applicable law, including the international obligations of the United States" (Arts. 51–303).

[14] This quote has been attributed to Eric Anderson, Planetary Resources CEO.

[15] This is less surprising when looking at the country's extremely solid fundamentals and its significant track record in the satellite sector dating back to the 1980s. When satellites ceased to be solely a government business and the market opened to private operators, Luxembourg was among the first movers in 1985 with Société Européenne des Satellites (SES).

[16] Loi du 20 juillet 2017 sur l'exploration et l'utilisation des ressources de l'espace. 20 July 2017.

[17] Art 2: "*Aucune personne ne peut explorer ou utiliser les ressources de l'espace sans être en possession d'un agrément de mission écrit du ou des ministres ayant dans leurs attributions l'économie et les activités de l'espace*".

obtaining the necessary licences, one[18] of which demands the incorporation of space companies in Luxembourg, which resulted in today's company hub effect we already mentioned.

This regulation was the result of an aggressive investment policy in the space sector, which led in 2016 to the creation of the Space Resource Initiative, attracting many foreign companies, including the already cited Deep Space Industries and Planetary Resources, with state funds allocated for feasibility studies. In some cases, the state even acquired shares in these companies.[19]

These laws have had the important effect of allowing the U.S. and Luxembourg to develop an ecosystem of companies specialized in space mining research, giving them an advantage over other countries, including Italy,[20] which still does not offer legal certainty in this regard. Others, like the UEA, have followed up on the same path, while more countries (e.g., France, Germany) are now looking at a comprehensive reorganisation of the space sector regulatory framework in a 'commercial' focus.

To date, and regardless of the national regulations, all countries remain, however, more or less explicitly anchored to the principles of the Outer Space Treaty of 1967, leaving the well-known wide margins of interpretation regarding the fate of extracted resources, and leaving the sector exposed to the geopolitical tensions highlighted in Sect. 2.

This is why everybody is now watching with particular attention the Artemis Accords.

While distant from being a formal treaty[21] (they ar more of a series of non-binding multilateral agreements between the government of the United States and other governments worldwide) they[22] still establish principles of international cooperation for peaceful space exploration, carried through a series of planned Artemis missions.

Importantly, the space policy promoted by the Artemis Accords explicitly allows private entities from signatory countries to exploit extracted resources (as explicitly

---

[18] Art. 4: "*L'agrément pour une mission ne peut être accordé que si le demandeur est une société anonyme, une société en commandite par actions ou une société à responsabilité limitée de droit luxembourgeois ou une société européenne ayant son siège social au Luxembourg*".

[19] This was for instance the case of Planetary Resources, state-owned for a share of 28 million USD.

[20] At present, Italy does not have specific legislation on the subject, and the new law currently under review neither regulates nor even explicitly mentions space mining (unlike what it does for frequencies), which objectively seems like a missed opportunity to provide clarity and to place Italy on par with the countries that have already legislated on the matter.

[21] Section One states their non-binding characters, as '*to establish a common vision* via*a practical set of principles, guidelines, and best practice to enhance the governance of the civil exploration and use of outer space with the intention of advancing the Artemis Programs*'.

[22] They are based on the 1967 Outer Space Treaty (OST1967) and were proposed by NASA and the U.S. Department of State in 2020 and, at May 2025, they have been signed by 55 countries, including the United Kingdom, Luxembourg, UAE, France, Germany, Spain, Japan, and India, although the exclusion of China and Russia limits the effectiveness of certain dispositions. Moreover, some Artemis signatories, such as Brazil, have previously expressed critical observations regarding national space mining regulations, raising challenging questions concerning implementation (Von der Dunk, 2023).

stated in Sect. 10[23] of the Accords; Deplano, 2021; Mostehar, 2020), potentially incentivising investments in specific technologies related to cislunar orbit.

6 Too Big to Quantify? The Environmental Risks of Space Mining and the Sustainability Dimension

Space sustainability has emerged as one of the most sensitive areas of discussion and debate in the last five years, due to the pressure, the worsening of orbital conditions, and the need to avoid a 'tragedy of the commons' (Hardin, 1968) in outer space.[24]

The sustainability dimension has, of course, a far more specific content when talking about space mining, and not just because mining itself is an industry that comes with a high price in externalities on Earth as well: it is for the unique challenges it is going to present when taking place in outer space. It is not by chance that it represents the core of the conceptual framework presented in this article.

This is where the three dimensions fully weigh in, and they can play a dual role, positive and negative. In positive, by limiting the dangers and the externalities of space mining. But they can also add to the challenges, in a variety of ways.

First of all, the economic pressure.

Without taking into account environmental considerations, the efforts to make the business profitable can lead to risky and unsustainable extraction practices, and even accidents in orbits, with the effect of worsening the debris crisis and potentially provoking irreversible ecosystem damage (Farinella & Cordelli, 1991; Martin-Lawson et al, 2023; Undeseth et al., 2021).

In this perspective the link to the technical and logistical aspects is evident, as only security-enhanced technical advances can ensure they meet economic objectives while incorporating sustainability principles to minimise environmental impact (Mo et al, 2024; Pardini & Anselmo, 2018, 2021) through clean extraction methods and efficient re-entry modalities.

But to make sure the technical advances are applicable and applied with the respect of appropriate industrial standards, the legal framework is, again, crucial to make sure environmental protections and responsible practices are enforced by all

[23] Artemis Accords, Sect. 10: "*The Signatories note that the utilization of space resources can benefit humankind by providing critical support for safe and sustainable operations. 2. The Signatories emphasize that the extraction and utilization of space resources, including any recovery from the surface or subsurface of the Moon, Mars, comets, or asteroids, should be executed in a manner that complies with the Outer Space Treaty and in support of safe and sustainable space activities. The Signatories affirm that the extraction of space resources does not inherently constitute national appropriation under Article II of the Outer Space Treaty, and that contracts and other legal instruments relating to space resources should be consistent with that Treaty.*"

[24] The 'tragedy of the commons', first proposed by Haardin (1968), is a scenario where independent groups make interest-driven, rational choices, which, due to the limited viewpoint, still end up depleting shared, limited resources, damaging everybody's future. A well-known phenomenon, although there is still a matter of debate how applicable it is to outer space.

the space mining actors, and not only the ones from countries that have, as discussed in the previous section, equipped themselves with national laws on resources.

An early application of robust regulations and standards from the pioneers will make simpler a wider adoption by the follow-up companies, so to speak. The major potential issue is, of course, imbalance between the various dimensions, especially if the regulatory framework is not given the required priority.

An actual risk would be, for instance, overemphasising economic gains and short-termed exploitation possibilities, ignoring environmental and ethical consequences.

Likewise, prioritising technical capabilities without aligning with economic viability and legal compliance can lead to operational failures, with all their related debris-related risks, unsustainable practices, and, eventually, to worsening international tensions.

An effective, globally accepted space governance seems, once more, a condition *sine qua non* for the entire sector.

7 Looking Ahead

Everyone acknowledges that space offers immense value to Earth through technological progress, scientific breakthroughs, and economic potential, and that, in order to translate all this potential into reality, requires a vision[25] that is compatible with a long-term sustainability of the activities in the space environment. As discussed in this article, this entails achieving a balance among economic viability, technical innovation, and legal clarity to ensure an environmentally sustainable, effective, and peaceful development of space industry. However, perspectives do diverge significantly when translating this vision into action, as countries and private operators often have distinct and competing goals shaped by their individual priorities and strategic interests.

And if the space industry is by definition "high risk, high reward," the edges of the sector, and space mining *in primis*, embody all the challenges, risks, and opportunities the industry itself promises. The economic and technical hurdles are enormous. But so are the legal and political implications, which, if not properly addressed, risk unraveling all efforts to exploit this potential.

Space mining and the issue of property rights concerning the appropriation of resources extracted from celestial bodies represent areas of potential major friction, given that international treaties explicitly state that no one may "own" space or celestial objects. Countries such as the United States and Luxembourg have enacted laws to address this apparent contradiction by establishing rights of appropriation for extracted resources, without extending such rights to the mining sites themselves,

---

[25] It is essential that this 'vision' is as much as possible a shared one at a global level. A recent INMARSAT report highlighted the varied ways in which countries perceive space, with views shaped by factors such as nationality, demographics, and education levels (Inmarsat, 2020). Recognising and considering these diverse perspectives and aspirations will be crucial for achieving any form of effective and successful space governance, (Paladini, 2024).

whether these are Near-Earth Objects (NEOs, i.e., asteroids orbiting Earth) or the Moon.

While regulatory hurdles can be effectively facilitated by the introduction of this kind of domestic legal regulations, this comes with disadvantages. When regulations are directly applicable and enforceable, they generally do not extend beyond national boundaries. On the other hand, when only certain countries possess comprehensive and prescriptive regulatory frameworks, it can create barriers to fair competition, which would not occur if all participants were subject to similar rules, benefiting the sector as a whole in the long term.

There are other issues. As a cognate discipline like maritime law shows, diverse national legal frameworks often lead to practices such as "nationality planning" or the use of "flags of convenience." This situation is a concern for the space sector as well (Marboe, 2012), where only some nations have introduced ad hoc legislation to regulate economic and commercial activities. Moreover, even where such legislation exists, it does not necessarily cover all areas of interest comprehensively, leading to a dangerous legal void.

All these contradictions inevitably translate, sooner or later, into geopolitical tensions, which are destined to grow as fast as competition for resources. There is evidence this is happening, with adverse effects not only on space mining or the space sector, but the world itself.

Agreeing on new global space governance principles,[26] to preserve both space as a 'province of humanity' and access to it for future generations seems, indeed, the only viable way for a prosperous and equitable future.

References

Artemis Accord. (2020). Retrieved April 30, 2025, from https://www.nasa.gov/wp-content/uploads/2022/11/Artemis-Accords-signed-13Oct2020.pdf

Buzan, B., & Waever, O. (2003). *2003 Regions and powers: The structure of international security*. Cambridge University Press.

Creswell, J. W. (1994). Research design: Qualitative and quantitative approaches. Sage Publications.

Delcambre, L. M. L., Liddle, S. W., Pastor, O., & Storey, V. C. (2018). A reference framework for conceptual modelling. In *Proceedings of the ER 2018, volume 11157 of Lecture Notes in Computer Science* (pp. 27–42). Springer.

Deplano, R. (2021). The Artemis Accords: Evolution or revolution in international space law? *International and Comparative Law Quarterly, 70*(3), 799 ss.

Deudney, D. H., & Matthew, R. A. (1999). *Contest- ed grounds: Security and conflict in the new environmental politics*. Albany State University Press.

Dolman, E. (2002). *2002 Astropolitik: Classic geopolitics in the space age* (p. 15). Frank Cass.

[26] There have been many frameworks and solutions offered in this sense, from a scenario soft governance, which hypothesises a set of "piecemeal limited treaties" of the kind proposed by Hertzfeld and von der Dunk (2005) and ad hoc solutions case by case, including a Star Trek-like Federation (Molz, 2014) an International Code of Conduct for Outer Space Activities (UNOOSA, sd). See Paladini (2023) for a discussion of the scenarios.

Elliot, L. (2002). Regional environmental security: Pursuing a non-traditional approach. In T. Andrew, H. Tan, & J. D. Kenneth Boutin (Eds.), *Non-traditional security issues*. Institute of Defence and Strategic Studies.

ESA. (2017). Near-Earth Objects—NEO Segment. Retrieved November 30, 2024, from http://www.esa.int/Our_Activities/Operations/Space_Situational_Awareness/Near-Earth_Objects_-_NEO_Segment

ESA. (2025). ESA Report on the Space Economy 2025. Retrieved April 30, 2025, from https://space-economy.esa.int/documents/tJMabTj61KkdGVOtF6SKw6wGSxicen6ajUWamCG3.pdf

ESPI. (2025). *Space venture 2024: Global investment dynamics*. https://www.espi.eu/reports/space-venture-2024/

EU. (2016). Communication from the Commission to the European Parliament, the Council, the European Economic and Social Committee and the Committee of the Regions. Space Strategy for Europe (COM 2016/705 Final), Brussels. 2016

Farinella, P., & Cordelli, A. (1991). The proliferation of orbiting fragments: A simple mathematical model. *Science & Global Security, 2*(4), 365–378.

Global Security. (2005). Advanced missions cost model. Retrieved April 30, 2025, from https://www.globalsecurity.org/military/intro/reference/calc/AMCM.htm

Hardin, G. (1968). The tragedy of the commons. *Science, 162*(3859), 1243–1248.

Hostbeck, L. (2015). Space weapons' concepts and their international security implications. In K. Schrogl, P. L. Hays, J. Robinson, D. Moura, & C. Giannopapa (Eds.), *Hand- book of space security* (p. 970). Springer Media.

Hertzfeld, H. R., & von der Dunk, F. (2005). *Bringing space law into the commercial world: Property rights without sovereignty*. Space, Cyber, and Telecommunications Law Program Faculty Publications. Retrieved April 30, 2025, from https://digitalcommons.unl.edu/spacelaw/15/

INMARSAT Report. (2022). The value of space: Technological advancements and economic opportunities. Retrieved April 30, 2025, from https://www.inmarsat.com/en/insights/corporate/2022/value-of-space.html

Jakhu, R. S., & Pelton, J. N. (2017). *Global space governance: An international study* (vol. 4). Springer International.

Keck Institute for Space Studies. (2012). *Asteroid retrieval feasibility study*. Retrieved April 30, 2025, from http://kiss.caltech.edu/final_reports/Asteroid_finalreport.pdf

Marboe (Ed.). (2012). *Soft law in outer space* (pp. 315–324). Böhlau.

Martin-Lawson, D., Paladini, S., Saha, K., & Yerushalmi, E. (2023). The cost of (Un)regulation: Shrinking Earth's orbits and the need for sustainable space governance. *Journal of Environmental Management, 349*, Article 119382.

Metzger, P. T., & Autry, G. W. (2022). *The cost of lunar landing pads with a trade study of construction methods*. arXiv. Retrieved May 30, 2025, from https://arxiv.org/abs/2205.00378

Mo, W., Kinsey, B., Vickers, J. et al. (2024). Conceptualizing space environmental sustainability. *npj Advanced Manufacturing, 1*, 2. Retrieved April 30, 2025, from https://doi.org/10.1038/s44334-024-00002-z

Moltz, J. (2014). *Crowded orbits: Conflict and cooperation in space*. Columbia University Press.

Mosteshar, S. (2020). Artemis: The discordant accords. *Journal of Space Law», 44*(2), 591 ss.

Mylopoulos, J. (2020). Philosophical foundations of conceptual modelling: What is a conceptual model? *ER Online Summer Seminars*. Retrieved April 30, 2025, from https://eross2020.inf.unibz.it/wp-content/uploads/2020/07/UniBZ-John.pdf

NASA Science Editorial Team. (2023). OSIRIS-REx delivers NASA's first asteroid sample to Earth. *NASA Science*, 24 September. Retrieved May 30, 2025, from https://science.nasa.gov/missions/osiris-rex/osirisrex-delivers-nasas-first-asteroid-sample-to-earth/

NASA. (2025). *Hayabusa2 Mission to Asteroid 162173 Ryugu*. Astromaterials Research and Exploration Science (ARES), NASA Johnson Space Center. Retrieved May 30, 20205, from https://ares.jsc.nasa.gov/missions/Hayabusa2/

Novaspace. Space Economy Report, 11th Ed 2025. Retrieved April 30, 2025, from https://digital-platform.euroconsult-ec.com/product/space-economy-report/

OECD. (2014). *The space economy at a glance 2014*. OECD.
OECD. (2019). *The space economy in figures: How space contributes to the global economy*. OECD Publishing. https://doi.org/10.1787/c5996201-en
Osburg, J., & Lee, M. (2022). Governance in space: Mining the Moon and beyond. *RAND Corporation*, 18 November. Retrieved May 30, 2025, from https://www.rand.org/pubs/commentary/2022/11/governance-in-space-mining-the-moon-and-beyond.html
Paladini, S. (2019). *The new frontiers of space*. Springer International Publishing.
Paladini, S., & Castellucci, I. (2022). *Sovereign states, private actors, and (national) space laws*. Oxford University Press.
Paladini, S. (2023). Unsustainable wars? The use of weapons in lower earth orbit. In D. Gruyter (Ed.), *Environmental injustice and catastrophe: How global insecurities threaten the future of humanity*.
Pardini, C., & Anselmo, L. (2018). Evaluating the environmental criticality of massive objects in LEO for debris mitigation and remediation. *Acta Astronautica, 145*, 51–75.
Pardini, C., & Anselmo, L. (2021). Evaluating the impact of space activities in low earth orbit. *Acta Astronautica, 184*, 11–22.
The Planetary Society. (2023). Cost of OSIRIS-REx. *The Planetary Society*. Retrieved May 30, 2025, from https://www.planetary.org/space-policy/cost-of-osiris-rex
Pfalzgraff, R. (2015). *International relations theory and space power*. Institute for National Strategic Studies.
Robinson, S., Arbez, G., Birta, L. G., Tolk, A., & Wagner, G. (2015, December). Conceptual modeling: Definition, purpose and benefits. In *2015 Winter Simulation Conference (WSC)* (pp. 2812–2826). IEEE.
Sheehan, M. (2015). Defining space security. In K.-U. Schrogl et al. (Eds.), *Handbook of space security*. Springer.
Schrogl, K., Hays, P. L., Robinson, J., Moura, D., & Giannopapa, C. (Eds.) (2015). *Handbook of space security* (vol. 14). Springer Media.
Smith, N. (2019). Giant asteroid has gold worth $700 quintillion. But it won't make us richer. *ThePrint*, 9 July. Retrieved May 30, 2025, from https://theprint.in/opinion/giant-asteroid-has-gold-worth-700-quintillion-but-it-wont-make-us-richer/260482/
The Space Foundation. (2024). The Space Report. Retrieved April 30, 2025, from https://www.thespacereport.org/
Space.com. (2012). Asteroid Mining Venture Backed by Google Execs, James Cameron Unveiled. Retrieved April 30, 2025, from https://www.space.com/15395-asteroidmining-planetary-resources.html
Space.com. (2023). Metal asteroid Psyche has a ridiculously high 'value.' But what does that even mean? 10 October. Retrieved May 30, 2025, from https://www.space.com/psyche-metal-asteroid-composition
SpaceNews. (2019a). Deep Space Industries acquired by Bradford Space. *SpaceNews*, 2 January. Retrieved May 10, 2025, from https://spacenews.com/deep-space-industries-acquired-by-bradford-space/
SpaceNews. (2019b). Asteroid mining company Planetary Resources acquired by blockchain firm. *SpaceNews*, 31 October. Retrieved May 30, 2025, from https://spacenews.com/asteroid-mining-company-planetary-resources-acquired-by-blockchain-firm/
SpaceNews. (2020). NASA selects four companies for lunar sample purchases. 3 December. Retrieved May 30, 2025, from https://spacenews.com/nasa-selects-four-companies-for-lunar-sample-purchases/
Thalheim, B. (2018). Conceptual model notions–a matter of controversy: Conceptual modelling and its lacunas. *Enterprise Modelling and Information Systems Architectures, 13*, 9–27.
Undseth, M., Jolly, C., & Olivari, M. (2021). The economics of space debris in perspective. Retrieved April 30, 2025, from https://conference.sdo.esoc.esa.int/proceedings/sdc8/paper/12/SDC8-paper12.pdf

United Nations Committee on the Peaceful Uses of Outer Space (UN COPUOS). (2018). Sustainability guidelines.

UNOOSA. (2022). Space supporting the sustainable development goals. Retrieved April 30, 2025, from https://www.unoosa.org/oosa/en/ourwork/space4sdgs/index.html

Valuables Consortium. (2021). Valuables consortium portal, Resources for the Future. Retrieved April 30, 2025, from https://www.rff.org/valuables/

von der Dunk, F. G. (2023). Property rights over the moon or on the moon? The Legality of space resource exploitation on celestial bodies. *Journal of Law & Innovation*, *6*(1), Article 3.

Wall, M. (2024). Space mining startup AstroForge aims to launch historic asteroid-landing mission in 2025. *Space.com*, 21 August. Retrieved May 30, 2025, from https://www.space.com/asteroid-mining-astroforge-docking-mission-2025

Warren, H., 2023. NASA, SpaceX launch Psyche mission to metallic asteroid. *NASASpaceFlight.com*, 12 October. Retrieved May 30, 2025, from https://www.nasaspaceflight.com/2023/10/psyche-launch/

Yahoo Finance, 2023. NASA administrator: China's space ambitions pose threat of land grab on the Moon. *Yahoo Finance*, 2 January. Retrieved May 30, 2025, from https://finance.yahoo.com/news/nasa-administrator-chinas-space-ambitions-pose-threat-of-land-grab-on-the-moon-100011667.html

Yarlagadda, S. (2022). Economics of the stars: The future of asteroid mining and the global economy. *Harvard International Review*, 8 April. Retrieved May 30, 2025, from https://hir.harvard.edu/economics-of-the-stars/

Important Aspects when Designing a Mining Tax in Outer Space

Pavel Semerád, Marian Dobranschi, Piotr Luty, and Petr Zimčík

Abstract As humanity works hard to develop tools which enable humans to extract mineral resources in outer space, we have focused in this chapter on the taxation of such activity because we believe that taxing is the only way how to safeguard mining while keeping peace and harmony with any extra-terrestrial civilisations. Concerning the taxation design, we also concentrated on the theories that we consider important aspects in terms of future taxation and assume step by step, i.e. the Fermi paradox, the Drake equation, the Kardashev scale, the planet Earth's inflation, the US dollar value and the Laffer curve. In our opinion, it is necessary to employ the accounting principle of prudence when exploring outer space since journeys through the universe may intersect with extra-terrestrial civilisations. Should that happen, such civilisations might demand compensation for the property benefits which humans had achieved within their territories. Hence we propose to incorporate this theoretical possibility as early as the taxation design stage, which will lead to a greater level of security when negotiating the allocation of outer space for the future. As a result, we consider, inter alia, three types of space civilisations, extending Semerád's proposal to include a proportional share as part of the mining tax and subsequent enforcement of compliance.

Keywords Asteroid mining · Drake equation · Fermi paradox · Kardashev scale · Mining taxation

---

P. Semerád (✉)
Sting University, Brno, Czech Republic
e-mail: pavel.semerad@sting.cz

M. Dobranschi
"Nicolae Balcescu" Land Forces Academy, Sibiu, Romania
e-mail: dobranschi.marian@armyacademy.ro

P. Luty
Wroclaw University of Economics and Business, Wrocław, Poland
e-mail: piotr.luty@ue.wroc.pl

P. Zimčík
NEWTON University, Prague, Czech Republic
e-mail: petr.zimcik@newton.university

H. Jahankhani et al. (eds.), *Space Mining*, Space Law and Policy,
https://doi.org/10.1007/978-981-95-5543-7_4

1 Introduction

It is the free will of every human whether or not they are willing to admit that life may exist anywhere in outer space and with it, as a pure theory, a civilisation at the same or even higher level of development as that on our planet (Determann & Malik, 2024; Döbler et al., 2023; Sagan & Drake, 1975). Likewise, it is absolutely legitimate when a human will argue that humanity is alone in outer space despite the existence of the enormous amount of stars and planets (Tipler, 1980). A judgement like this is particularly complicated for religious individuals since this inevitably creates a conflict in how they perceive God (Delio, 2007).

Although even a minimal probability of the existence of an extra-terrestrial civilisation may be difficult to accept without evidence, humanity has been trying to make contact with such civilisation for more than 100 years. Cases of intercepting signals from Mars which were accomplished by e.g. Guglielmo Marconi (Leigh Hester, 2018) and Nikola Tesla (Richmond Times, 1901) at the beginning of the 20th century are well known. It was a partial success, however, as these probably heard just some noise emitted from Jupiter (Corum & Corum, 1996).

The first more recent—albeit unsuccessful—attempts to detect radio signals from outer space date back to the 1960s and are associated with Frank Drake, the US physicist. Drake entered history by, among other things, introducing his equation in 1961 (Golden, 2021); the equation allows one to determine, in theory, the number of extra-terrestrial civilisations existing at the same moment and capable of communication (*1*).

The Drake equation is (SETI Institute, 2024):

$$N = R_{*} \times F_p \times n_e \times f_1 \times f_i \times f_c \times L \tag{1}$$

where

- N The number of civilizations in the Milky Way galaxy whose electromagnetic emissions are detectable.
- R_{*} The rate of formation of stars suitable for the development of intelligent life (number per year).
- f_{p} The fraction of those stars with planetary systems.
- n_{e} The number of planets, per solar system, with an environment suitable for life.
- f_{l} The fraction of suitable planets on which life actually appears.
- f_{i} The fraction of life-bearing planets on which intelligent life emerges.
- f_{c} The fraction of civilizations that develop a technology that produces detectable signs of their existence.
- L The average length of time such civilizations produce such signs (years).

Based on probability theory, Drake triggered the discussion that there may be life in the universe. For instance, Frank and Sullivan (Frank & Sullivan, 2016) assume that if "*a habitable zone planet develops a technological species larger than* $\sim 10^{-24}$, *humanity is not the only time technological intelligence has evolved*."

It might be the reason why Enrico Fermi, the Italian physicist, asked: "*Where is everybody?*", which is referred to as the Fermi paradox today (DeVito, 2024). Here, arguing is possible, as e.g. Gray (2015) actually does, whether or not Hart (Hart, 1975), the American astrophysicist, is the father of the idea with his statement: "*They are not here—hence they do not exist.*" Hart (1975) also states that "*there are no intelligent beings from outer space now on Earth. There may have been visitors in the past, but none of them settled or colonised the planet.*" This way Hart's argument not only does not refute extra-terrestrial life, but even admits it.

Carl Sagan, the American astronomer, took a similar approach to the topic, stating: "The *universe* is a *pretty big place. If it's just us, seems like an awful waste* of *space*" (Vincent, 2023). Hence Sagan participated, in 1977, in transmitting audio-visual content—referred to as the Golden Record—from Earth. Carrying information about the planet and life on it, the message is assumed to be already beyond our solar system. It includes information on the location of the Earth and instructions on how to access and read the content. Albeit it is a significant technological achievement, there have been criticisms that this content could be confusing to potential recipients (Traphagan, 2021).

However, efforts to make contact continued (Atri et al., 2011). One of history's greatest moments, most hopeful for those attempting to get in touch with extra-terrestrial civilisations, was the Wow! signal of 15 August 1977 (Gray & Marvel, 2001). That night, the radio telescope at Ohio University intercepted a signal that, when transcribed, contained numbers and letters.

Similar activities and experiments with content and reaching out can be traced to the present day—and it is unlikely that humanity will stop them. The desire to discover life in outer space is so enormous that it is not only motivating humanity to develop new technologies; indeed, the time is near when we will begin to colonise the universe. Mars is assumed to become the first target after 2030 (Szocik et al., 2020). The 20,000 near-Earth asteroids may become objects of interest as well (Xie et al., 2021).

This takes us close to a historic milestone when we, as mankind, start to populate other celestial bodies. In this regard, the increasing number of confirmed exoplanets, i.e. planets that orbit a star other than our Sun. NASA (2024) has already confirmed 5,566 exoplanets and knows of 10,059 more identified objects that could have such a status. In total, there are 4,145 known stars with their own planets that form planetary systems. These are all celestial bodies where we could search for and find life, or, the conditions for life to form.

2 Why we Have Not Yet Been Able to Make Contact with Extra-Terrestrial Civilisations

Let us return to the Fermi paradox as such. To explain why we have so far failed to get in touch with an extra-terrestrial civilisation, we have several scenarios to handle:

(1) The pessimistic scenarios suggest that either extra-terrestrial life does not exist in the universe or life is so rare that even if it could emerge, it would (immediately) spontaneously cease to exist (Haqq-Misra & Baum, 2009). If that is the case, Earth would be the only planet where life could emerge and evolve. Given the number of discovered planets, this assumption seems rather unlikely.
(2) A much more likely explanation is that life may have emerged, but is still at the level of primordial organisms; alternatively, such creatures are insufficiently developed in terms of intelligence and unaware of themselves (Gallup & Faliveno, 2022).
(3) The lack of interest in making contact with us may also be due to extra-terrestrial civilisations fearing for their safety. Perhaps they are as worried as we are about how further collaboration would evolve (Michael, 2014) since, after all, we cannot rule out the fact that after some time one group might want to exterminate the other. Diseases as well as invading unwanted new species (lacking natural controls) could affect life as well, irreversibly damaging the planet's biosystem. After all, we struggle with our own invasive species on planet Earth (Monceau et al., 2014).
(4) Another assumption may involve the fact that there is a large distance between Earth and other inhabited planets. Considering that it could be (hundreds of) thousands of light years, then it will be a long time (*2*) before we can establish a dialogue (D_T).

$$D_T = T_1 + R_T + T_2 \tag{2}$$

The message sent would actually have to travel the distance from Earth to any inhabited body (T_1). The recipients would have to be able to decode the message after receiving and reply to it (R_T). Subsequently, their response would have to travel the entire distance back to planet Earth (T_2). In theory, the response could be replaced by some kind of signal that would be possible to detect from planet Earth.

Given such circumstances, it may also be that the recipient (or sender) may not live to see the reply, as civilisation may have ceased by then. An example of this could be the collision of planet Earth with another body, as was the case 66 million years ago when the dinosaurs died out after an asteroid impact (Black, 2022). Such events may be recurring at regular intervals. In such a case, it seems to be easy to understand that the universe is tremendously silent.

(5) It can also be at least admitted that extra-terrestrial civilisations already know about us and are watching us. If this is the case, they would be more technologically advanced than us, since we have not yet detected them. This is probably because they do not wish to threaten peace and security in outer space, or they do not want to disrupt human evolution and consider us an experiment. This is often referred to as the Zoo Hypothesis (Crawford & Schulze-Makuch, 2023; de Magalhães, 2016).

If extra-terrestrial civilisations have been observing us for extended periods, then they have witnessed and still are witnessing many experiments in which humans have tried to murder each other (e.g. World Wars I and II). They may also have observed the weapons we have used against each other (e.g. dropping the atomic bombs on Nagasaki and Hiroshima). If humanity is capable of doing this against its own species, it might do the same against other civilisations. The question is also whether the beings with whom we will get in touch would not act similarly towards ourselves (Kecskes, 2002; Santana, 2021).

3 How we can Measure how much a Civilisation is Advanced

As we look for and assess the level of advancement of extra-terrestrial civilisations, it is essential to mention the Kardashev scale (Zhang et al., 2023). A Soviet astrophysicist, Nikolai Semyonovich Kardashev defined the level of advancement of civilisations in the universe, based on the observation of the exponential increase in energy consumption, as follows (Kardashev, 1964):

- Type 1 civilisations can make use of all the resources only on their home planets.
- Type 2 civilisations can make use of all the energy from their nearest stars.
- Type 3 civilisations can make use of all the energy from their home galaxies.
- In theory, we could also consider a Type 4 civilisation that can make use of all the resources in the universe (Gray, 2020).

Based on the civilisation's level of advancement, the thermodynamic manifestations of energy consumption should be possible to observe outside the body where the consumption occurs. For example, lights on the Earth's surface are possible to observe at night as well, albeit our planet is currently at the first level of the Kardashev scale (c. 0.72).

Let us suppose that humans can extract minerals in outer space. We can assume that this would not infringe on the territory of another civilisation in our solar system. However, there is a risk of (future) conflict with a Type 3 civilisation, which could have fatal consequences for us.

For this reason, the steps we take as we colonise and explore outer space should be more cautious, meaning to avoid plundering and appropriating the wealth of the universe for the benefit of planet Earth alone, as another civilisation could claim the wealth, too. Although contact with extra-terrestrial civilisations may never occur, humanity should be prepared to face a situation of the benefits of space mining being claimed at any time in the future. From this perspective, humanity needs to take the Fermi paradox into account when planning extra-terrestrial activities.

4 How to Keep the Peace in Outer Space

While ensuring safety in terms of clashes with celestial bodies (Schmidt, 2024) and during human colonisation of outer space should be given absolute priority since life on Earth could become extinct if these efforts fail, there is one crucial difference. Mathematical models can be used to estimate the probability and period of the threat of such a clash if tracking the trajectory of celestial bodies; hence a great deal of attention this topic has received in recent decades (Alvarez et al., 1980).

Yet models for encounters with extra-terrestrial civilisations do not exist. Since we can neither confirm nor refute the existence of extra-terrestrial civilisations, we should follow the accounting principle of prudence (Hansson, 2020; Schipper, 2022) demanding humanity to rather assume the worse scenario and prepare for future compensation of property benefits in handling resources in outer space.

If humanity is concerned about safe and peaceful extraction in space, this means that it must use a tool that enables redistribution of the property benefit among all potential groups claiming any share. These are (*i*) individual states on Earth; (*ii*) collaborating colonies; and (*iii*) extra-terrestrial civilisations within their claimed ownership in the solar system. The instrument could involve a space mining tax (Semerád, 2023). No matter how large the share we decide to redistribute to each group, it will be important to keep the value of that share for possible future transactions.

The biggest barrier to future transaction is inflation (Ha et al., 2022; Modigliani, 1979; Wang et al., 2022). Hence, inflation will not make it possible to keep the value for settlement with extra-terrestrial civilisations in terrestrial money. As can be seen in Fig. 1, the value of the US dollar has declined significantly from 1635 to 2020 and will be even lower in 1,000 years. Looking at a horizon of 10,000 years and beyond, it will even be negligible.

Compared to other commodities such as gold, the US dollar is less stable and less likely to retain its value (Zagaglia & Marzo, 2022), as its exchange rate is difficult to estimate over the long term (Engel & Wu, 2023). This is also why there is talk of a global decline in the importance of the US dollar and its possible replacement by a different currency, i.e. the Yuan (Siddiqui, 2020; Sielska, 2020). Future exchange rates could harm the value being maintained.

Another issue we face is the value of money itself on our planet since money only has value if this value is recognised by the recipient of the transaction. However, money could be considered totally worthless by extra-terrestrial civilisations. As a result, keeping any compensation for extra-terrestrial civilisations in Earth money is utopian from the security aspect. In addition, there would be a danger of future wars being fought over money immediately on Earth, which could lead to a Great Filter situation right on our planet.

Compensation must therefore be kept on neutral ground outside our planet, in a form that retains its value with minimal inflation. Preserving the compensation in the value of the extracted mineral material therefore seems to be the most appropriate option. Semerád (2023) favours this solution as well. However, the author does not

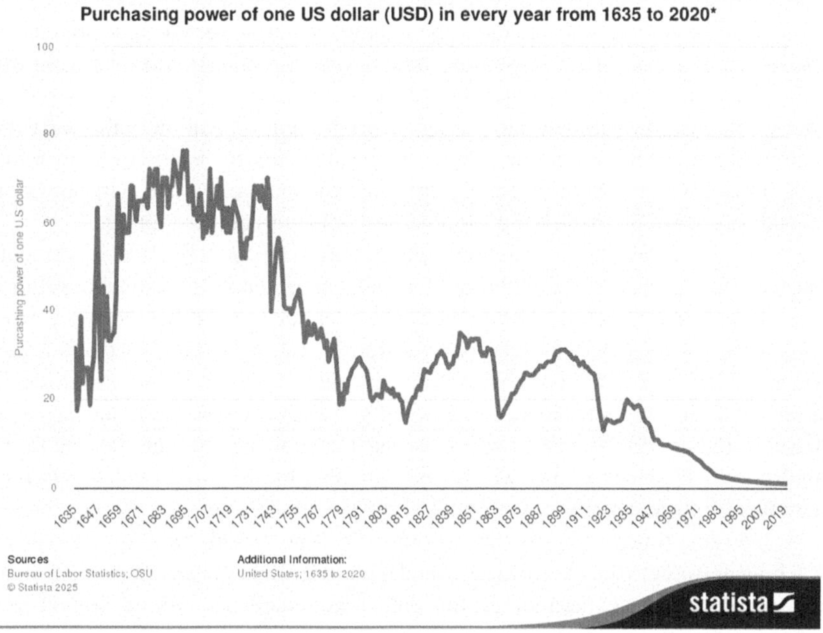

Fig. 1 Purchasing power of one US dollar (USD) from 1635 to 2020. *Source* O'Neill (2022)

admit the threat of multiple civilisations declaring the claim either simultaneously or over time.

The point is that if only the first civilisation were to appropriate the compensation, this would lead to a stalemate since each additional claim would not be granted. It is therefore necessary to specify rules for the distribution of the minerals extracted based on the stage of mining.

5 Mining Stages

5.1 Stage 1: Mining Within Our Solar System

It is clear that if humanity starts extraction in outer space, it will first occur in our solar system (Halidu et al., 2024; Lee & Fong, 2024; Palaszewski, 2009). We have explored this space rather extensively and so it is believed that we are alone within it. For this reason, let us keep 10% of the value of the minerals extracted to be distributed among the states on planet Earth. This sharing would take the form of a space tax and would present a necessary condition for compensating the less developed (and non-space-mining) states. Otherwise, this could trigger wars over compensation.

We consider Stage 1 to involve mining operations extending up to the boundaries of our solar system. In this situation, the inhabitants on planet Earth—and, later on, perhaps a colony on Mars—would become a Type 2 civilisation on the Kardashev scale. No mining share is applied to a Type 3 civilisation. This is because of the need for massive investment into our space programmes; taxation in the early days could limit the incentive for our growth. The cost of mining itself will be extremely high, although technology will make it more efficient—e.g. there will not be any unnecessary waste in identifying suitable/unsuitable targets for mining.

For the development of extraction technologies to be cost-effective, it is essential that the costs are recovered and that our civilisation makes a reasonable profit (in terms of imported minerals). Otherwise, the Laffer curve theory would apply, where from a certain point (Laffer point) taxpayers refuse any further taxation (Laffer, 1981; Wanniski, 1978). When too high, taxation may either lead to resistance to invest any further or fraud will start to occur, leading to lower tax revenue (Clausing et al., 2020; Clotfelter, 1983). Neither option is sustainable in the long run. For the Earth's civilisation, off-planet mining will be the only possible way to provide enough raw materials for the economy to function (Northey et al., 2014).

At the same time, we apply the option that a Type 3 civilisation may already be watching us (refer to the Zoo scenario), and therefore might not mind that we continue to evolve during this experiment. If this option were realistic, nothing would protect our planet anyway, even if we left some compensation in place at this stage.

5.2 *Stage 2: Building Colonies and Mining Outside Our Solar System*

Important milestones in space mining will involve (*i*) forming the first colony; and (*ii*) crossing the boundaries of our solar system. We mentioned previously that the planet Mars may become the first colony (Levchenko et al., 2021; Zubrin, 2019). As long as the development of such a colony is funded and supplied from Earth, such a colony should not receive property benefits within a space mining tax. Stephen Hawking (Sample, 2016), too, discussed the fact that colonies would not be self-sustaining in the next 100 years at least.

Once human civilisation has crossed the boundaries of the solar system, more colonies can be assumed to be strategically set up, while increasing the likelihood that our activities intrude into another civilisation's space.

The same rule of thumb applies to those additional colonies as it does to building a colony on Mars. The only difference will be that the equipment for these colonies will mostly use materials extracted from outer space (Grandl, 2017) since imports from Earth would be disproportionately expensive concerning costly transportation.

As long as Earthlings and possibly the Mars population cooperate with the colonies, the sharing will be based on the merits of the colonies' viability. Consequently, the colonies are assumed to pay Earth by giving up future claims on the extracted resources in their surroundings.

If however, the paths of Earth's inhabitants and those of colonies diverge and the colonies become independent, then it will be necessary to address the sharing of the profits from space extraction. At this point, Semerád (2023) proposes a sharing ratio of 10% in favour of planet Earth (E) and 10% in favour of the colony (C_o). As this is a problematic issue, however, as Earth-Mars collaboration can be assumed to persist within the solar system, our proposal for amendment is that no more than 20% tax be collected within the solar system and that it be redistributed equally as per the number of residents within the solar system ($E+C_0$). However, this will be applicable upon a precondition that Mars (or another colony) can provide for its own needs independently of planet Earth (Szocik et al., 2016).

At this point, the tax will be paid by the party that delivers the extracted resources to its own celestial body. Any mutual set-off of claims will be possible. To avoid fraud in settlements and imports, it is necessary to set up rules for the supervision of extraction. Mining should be done based on uniform rules for the valuation of benefits and under the supervision of an independent organisation; this should not be left to the will of the states themselves.

Mining outside our solar system in outer space carries with it the potential threat of a clash with a civilisation that has not emerged in our solar system. To pay future benefits to this potential civilisation, refunds worth 10% of the extracted resources will be allocated ($C_{1\text{-}2}$). This will increase the amount of the tax (*3*) to 30% after accounting for the share for the previous two groups. However, this will only offset the claim of the civilisation that inhabits the respective solar system.

$$Tax = (E + C_0) + C_{1-2} + C_3 \tag{3}$$

where

E Earth,
C_0 Colonies,
$C_{1\text{-}2}$ Civilisation C_1/C_2 (10%),
C_3 Type 3 Civilisation.

Since our proposal, however, also allows for the fact that another civilisation, e.g. Type 3 as per the Kardashev scale, will also lay a claim, the share of the potential settlement will be increased by another 10% (C_3). This will increase the total tax outside the solar system to 40% (for comparison, 20% is the proposal within our solar system). Given the fact that civilisations can claim the entitlement independently at different time horizons, the compensation should not be collected in one place. The two 10% compensations should be kept separate from each other and as close as possible to the extraction site. This eliminates transportation costs from the bodies being extracted.

Although we are aware that even after two claims are settled, there may be others, it is necessary to consider the Drake equation, which assumes that there are only a certain number of co-existing and communicating extra-terrestrial civilisations at any one time. Therefore, our taxation of 40% is final and is a maximum limit since we can assume that new future civilisations will not claim the benefit after millions and billions of years retroactively.

6 Conclusion

A period is near when we will have to begin extracting minerals on celestial bodies other than Earth. The reasons are simple: (*i*) terrestrial sites are gradually dwindling, and (*ii*) people want to find out what it looks like in outer space and beyond our solar system.

Regardless of whether or not people of the Earth believe in extra-terrestrial life, they should keep the accounting principle of prudence in mind when appropriating property benefits in the universe. It is therefore necessary to establish rules for our peaceful exploration and exploitation of resources in outer space, meaning that we should generate enough resources to cover claims for compensation for the benefit of other civilisations at any time in the future. In our chapter, we assume civilisations based on the Kardashev scale, i.e. Type 1, Type 2 and Type 3 civilisations, for as Fermi (Fermi paradox), Drake (Drake equation) and Hart state, we cannot fully rule out that we are not alone in the universe. Encountering an extra-terrestrial civilisation, if any, however tempting for us, could have fatal consequences for life on Earth. To exist in peace and harmony in the universe in the long run, it is necessary to take into account the potential contact that could occur at any time in the future.

It is also partly why it is not advisable to keep any compensation in Earth currencies as these may lose value and become the subject of potential wars on Earth. It is preferable to retain a share of the mineral extracted right on the body being extracted.

We propose to use the share when redistributing benefits as follows: 10% for Earth states, 10% for colonies, 10% for civilisations outside our solar system and 10% for Type 3 civilisations. For the beginning, however, we suggest that taxation should not exceed 20%, as higher taxation could discourage Earthlings from activity (parallel to the Laffer curve) while proposing an increase (up to 40%) when our technological level allows us to mine beyond our solar system.

References

Alvarez, L. W., Alvarez, W., Asaro, F., & Michel, H. V. (1980). Extraterrestrial cause for the cretaceous-tertiary extinction. *Science, 208*(4448), 1095–1108. https://doi.org/10.1126/science.208.4448.1095

Atri, D., DeMarines, J., & Haqq-Misra, J. (2011). A protocol for messaging to extraterrestrial intelligence. *Space Policy, 27*(3), 165–169. https://doi.org/10.1016/j.spacepol.2011.01.001

Black, R. (2022). *The last days of the dinosaurs: An asteroid, extinction and the beginning of our world: Vol. First Edition*. St. Martin's Press.

Clausing, K. A., Saez, E., & Zucman, G. (2020). Ending corporate tax avoidance and tax competition: A plan to collect the tax deficit of multinationals. *SSRN Electronic Journal*. https://doi.org/10.2139/ssrn.3655850

Clotfelter, C. T. (1983). Tax evasion and tax rates: An analysis of individual returns. *The Review of Economics and Statistics, 65*(3), 363. https://doi.org/10.2307/1924181

Corum, K. L., & Corum, J. F. (1996). Nikola Tesla and the planetary radio signals. *Nikola Tesla and The Electrical Signals of Planetary Origin*.

Crawford, I. A., & Schulze-Makuch, D. (2023). Is the apparent absence of extraterrestrial technological civilizations down to the zoo hypothesis or nothing? *Nature Astronomy, 8*(1), 44–49. https://doi.org/10.1038/s41550-023-02134-2

de Magalhães, J. P. (2016). A direct communication proposal to test the Zoo Hypothesis. *Space Policy, 38*, 22–26. https://doi.org/10.1016/j.spacepol.2016.06.001

Delio, I. (2007). Christ and extraterrestrial life. *Theology and Science, 5*(3), 249–265. https://doi.org/10.1080/14746700701622008

Determann, J. M., & Malik, S. A. (2024). *Islamic theology and extraterrestrial life: New frontiers in science and religion*. Bloomsbury Publishing.

DeVito, C. L. (2024). The Fermi paradox and the Drake equation. *Acta Astronautica, 215*, 51–53. https://doi.org/10.1016/j.actaastro.2023.11.036

Döbler, N. A., El Amri, Y., & Carbon, C.-C. (2023). Is there anybody out there? Can individual loneliness, need for closure, and religiosity predict the belief in extraterrestrial life and intelligence? *Discover Psychology, 3*(1), 21. https://doi.org/10.1007/s44202-023-00076-4

Engel, C., & Wu, S. P. Y. (2023). Forecasting the U.S. Dollar in the 21st century. *Journal of International Economics, 141*, 103715. https://doi.org/10.1016/j.jinteco.2023.103715

Frank, A., & Sullivan, W. T. (2016). A new empirical constraint on the prevalence of technological species in the universe. *Astrobiology, 16*(5), 359–362. https://doi.org/10.1089/ast.2015.1418

Gallup, G. G., Jr., & Faliveno, H. E. (2022). Extraterrestrial intelligence: A cognitive evolutionary perspective. *Journal of Astrobiology, 12*, 65–75.

Golden, L. M. (2021). A joint mind consideration of the Drake equation in the search for extraterrestrial intelligence. *Acta Astronautica, 185*, 333–336. https://doi.org/10.1016/j.actaastro.2021.03.020

Grandl, W. (2017). Human life in the solar system. *REACH, 5*, 9–21. https://doi.org/10.1016/j.reach.2017.03.001

Gray, R. H. (2015). The fermi paradox is neither fermi's nor a paradox. *Astrobiology, 15*(3), 195–199. https://doi.org/10.1089/ast.2014.1247

Gray, R. H. (2020). The extended Kardashev scale. *The Astronomical Journal, 159*(5), 228. https://doi.org/10.3847/1538-3881/ab792b

Gray, R. H., & Marvel, K. B. (2001). A VLA search for the Ohio state "Wow." *The Astrophysical Journal, 546*(2), 1171–1177. https://doi.org/10.1086/318272

Ha, J., Kose, M. A., & Ohnsorge, F. (2022). From low to high inflation: Implications for emerging market and developing economies. *SSRN Electronic Journal*. https://doi.org/10.2139/ssrn.4074459

Halidu, O. B., Amidu, M., Asiedu, E., & Asare, N. (2024). *Effective and sustainable management of natural resources in Africa: The roles of environmental tax and public participation* (pp. 175–223). https://doi.org/10.1007/978-3-031-58124-3_9

Hansson, S. O. (2020). How extreme is the precautionary principle? *NanoEthics, 14*(3), 245–257. https://doi.org/10.1007/s11569-020-00373-5
Haqq-Misra, J. D., & Baum, S. D. (2009). The sustainability solution to the fermi paradox. *Journal of the British Interplanetary Society.*
Hart, M. H. (1975). *Explanation for the Absence of Extraterrestrials on Earth, 16*, 128.
Kardashev, N. S. (1964). *Transmission of information by extraterrestrial civilizations, 8*, 217.
Kecskes, C. (2002). Scenarios which may lead to the rise of an asteroid-based technical civilisation. *Acta Astronautica, 50*(9), 569–577. https://doi.org/10.1016/S0094-5765(01)00208-9
Laffer, A. B. (1981). Government exactions and revenue deficiencies. *Cato Journal, 1*(1), 1–21. https://EconPapers.repec.org/RePEc:cto:journl:v:1:y:1981:i:1:p:1-21
Lee, Y., & Fong, E. A. (2024). The art of living together: Space mining ecosystem, sustainability and accountability. *Accounting, Auditing & Accountability Journal.* https://doi.org/10.1108/AAAJ-12-2022-6174
Leigh Hester, J. (2018). *Everybody shut up! We're listening to mars.*
Levchenko, I., Xu, S., Mazouffre, S., Keidar, M., & Bazaka, K. (2021). Mars colonization: Beyond getting there. In *Terraforming mars* (pp. 73–98). Wiley. https://doi.org/10.1002/9781119761990.ch5
Michael, G. (2014). *Extraterrestrial intelligence: Friends, foes, or just curious?* (pp. 25–35). https://doi.org/10.1007/978-3-642-37750-1_2
Modigliani, F., & Cohn, R. A. (1979). Inflation, rational valuation and the market. *Financial Analysts Journal, 35*(2), 24–44.
Monceau, K., Bonnard, O., & Thiéry, D. (2014). Vespa velutina: A new invasive predator of honeybees in Europe. *Journal of Pest Science, 87*(1), 1–16. https://doi.org/10.1007/s10340-013-0537-3
NASA. (2024). *Exoplanet exploration.*
Northey, S., Mohr, S., Mudd, G. M., Weng, Z., & Giurco, D. (2014). Modelling future copper ore grade decline based on a detailed assessment of copper resources and mining. *Resources, Conservation and Recycling, 83*, 190–201. https://doi.org/10.1016/j.resconrec.2013.10.005
O'Neill, A. (2022). *Purchasing power of one US dollar (USD) in every year from 1635 to 2020*.*
Palaszewski, B. (2009, August 2). Atmospheric mining in the outer solar system: Mining design issues and considerations. In *45th AIAA/ASME/SAE/ASEE Joint Propulsion Conference & Exhibit.* https://doi.org/10.2514/6.2009-4961
Richmond Times. (1901). *Nikola Tesla promises communication with mars.*
Sagan, C., & Drake, F. (1975). The search for extraterrestrial intelligence. *Scientific American, 232*(5), 80–89. http://www.jstor.org/stable/24949801
Sample, I. (2016). *Most threats to humans come from science and technology, warns Hawking.*
Santana, C. (2021). We come in peace? A rational approach to METI. *Space Policy, 57*, Article 101430. https://doi.org/10.1016/j.spacepol.2021.101430
Schipper, K. (2022). Why do accounting issues end up in the 'too difficult' box? *Accounting and Business Research, 52*(5), 482–506. https://doi.org/10.1080/00014788.2022.2079686
Schmidt, N. (2024). Planetary defense governance: Thirty years of development and the multilateral future. *Acta Astronautica, 214*, 343–355. https://doi.org/10.1016/j.actaastro.2023.10.050
Semerád, P. (2023). Asteroid mining tax as a tool to keep peace in outer space. *Space Policy, 65*, Article 101555. https://doi.org/10.1016/j.spacepol.2023.101555
SETI Institute. (2024). *Drake equation.*
Siddiqui, K. (2020). The U.S. dollar and the world economy: A Critical Review. *Athens Journal of Business & Economics, 6*(1), 21–44. https://doi.org/10.30958/ajbe.6-1-2
Sielska, A. (2020). Changes in the international monetary system: opportunities and risks for the US dollar. *Economic Annals-XXI*, 5–6.
Szocik, K., Lysenko-Ryba, K., Banaś, S., & Mazur, S. (2016). Political and legal challenges in a Mars colony. *Space Policy, 38*, 27–29. https://doi.org/10.1016/j.spacepol.2016.05.012

Szocik, K., Wójtowicz, T., & Braddock, M. (2020). The martian: Possible scenarios for a future human society on mars. *Space Policy, 54*, Article 101388. https://doi.org/10.1016/j.spacepol.2020.101388

Tipler, F. J. (1980). *Extraterrestrial intelligent beings do not exist* (pp. 267–281). Royal Astronomical Society.

Traphagan, J. W. (2021). Should we lie to extraterrestrials? A critique of the voyager golden records. *Space Policy, 57*, Article 101440. https://doi.org/10.1016/j.spacepol.2021.101440

Vincent, L. (2023). *Starlight odyssey: A journey across the galaxy*. RWG Publishing.

Wang, G., Sharma, P., Jain, V., Shukla, A., Shahzad Shabbir, M., Tabash, M. I., & Chawla, C. (2022). The relationship among oil prices volatility, inflation rate, and sustainable economic growth: Evidence from top oil importer and exporter countries. *Resources Policy, 77*, Article 102674. https://doi.org/10.1016/j.resourpol.2022.102674

Wanniski, J. (1978). Taxes, revenues, and the Laffer curve. *The Public Interest, 50*(3).

Xie, R., Bennett, N. J., & Dempster, A. G. (2021). Target evaluation for near earth asteroid long-term mining missions. *Acta Astronautica, 181*, 249–270. https://doi.org/10.1016/j.actaastro.2021.01.011

Zagaglia, P., & Marzo, M. (2022). *Commodities* (2nd edition). Chapman and Hall/CRC.

Zhang, A., Yang, J., Luo, Y., & Fan, S. (2023). Forecasting the progression of human civilization on the Kardashev Scale through 2060 with a machine learning approach. *Scientific Reports, 13*(1), 11305. https://doi.org/10.1038/s41598-023-38351-y

Zubrin, R. (2019). Why we earthlings should colonize mars! *Theology and Science, 17*(3), 305–316. https://doi.org/10.1080/14746700.2019.1632519

UN-Outer Space Treaty The Magna Carta for Mining?

Sarah Jane Fox

Abstract The Outer Space Treaty (OST) has been recognised as the Magna Carta for international space, transpiring as a result of the space race between the United States and the then Soviet Union. This occurred at a time of global tension—in what is recognised as the 'Cold-War' period, whereby technology from World War II served to facilitate the movement and competition in space. This chapter sets out to explore the content of the OST in terms of the application to the new commercial pursuits and competition in space with specific focus on space resources and mining. Commentary is provided in terms of the scope of the Treaty and the permissibility and therefore acceptance and/or legalities of such related activities. Consideration is also given to the Moon Agreement and national directions. It finds that there is a symbiotic relationship of war and resources which will no doubt also transpire to a space domain; and that, the OST provides non-conclusive direction in terms of use and acquisition of resources, while it lays silent in terms of mining and the sharing of resources. Given this, and based on lessons from the Earth, it is predicated that, in time, ownership claims to resources, including mining, will lead to further disharmony and conflict, not only on Earth but in space.

Keywords United Nations · War · Global commons · Legalities · Astro-politics · United States

1 Introduction

The founding principles for the development and, therefore, governance of outer space, is the 1967 treaty entitled, the '*Treaty on Principles Governing the Activities of States in the Exploration and Use of Outer Space, including the Moon and Other Celestial Bodies*' (commonly referred to as the Outer Space Treaty—OST). This is generally regarded as the *Magna Carta* for international space law (Finch, 1984; Fox, 2024a; Qizhi, 1997).

S. J. Fox (✉)
University of Leicester, Leicester, UK
e-mail: sarah.fox@leicester.ac.uk

H. Jahankhani et al. (eds.), *Space Mining*, Space Law and Policy,
https://doi.org/10.1007/978-981-95-5543-7_5

Since then, there have been several other key treaties, which together are collectively referred to as the five United Nations treaties on outer space, these spanning the period 1967–1979. The last one being the 1979 Moon Agreement (MA) that entered into force on 11 July 1984. This said, the MA remains somewhat controversial and not widely accepted and endorsed and, hence, there are grounds to question whether, not just this Agreement, but, arguably, all of the treaties, have stood the test of time and, if they are fit for the current and future period, where there is likely to be a resurgence of manned exploration to other planets and bodies. The rapidly evolving and growing global space economy is expected to be worth approximately $1.8 trillion by 2035 (McKinsey & Company, 2024). Space is now highly commercialised. This new space renaissance period is also anticipated to encompass new actions and undertakings. This includes seeking resources which lie in space and inevitably mining ventures, which lead to questions being asked in terms of the scope of the treaties and the permissibility and therefore acceptance and/or legalities of such activities. In this chapter, consideration and discussion is largely given to the OST.

However, it is important to understand the background and history that led to both activities in space and also the role undertaken by the United Nations. Hence, this chapter seeks to, firstly, provide contextualisation and some key timelines that led to the OST, before passing commentary as to the contents and, hence, the relevance to space mining. It will explore whether, and to what degree, space mining is potentially overtly or covertly covered, and/or to the possible guidance or interpretation that could be formed or deduced from the contents of the OST. In so doing, any potential gaps will also be considered. Brief commentary will be provided as to the linkage to the development of nations, national approaches and the role and position of private companies. In summary, an overview of the current understandings and viewpoints will be factored in. The approach is based upon a legal and policy perspective which also overlaps into the related area of international relations and, hence, leads to astro-politics.

2 *In the Beginning ……*

The OST arguably played catch-up in terms of establishing principles for *activities* in space. Activities having been occurring since the end of the 1940s, when in 1949, the "Bumper-WAC" is said to have become the first human-made object to enter space, when it climbed to an altitude of 393 km (244 miles) (NASA, 2008).

The 1950s marked a flurry of events by both the then Soviet Union (further discussed below in terms of the position of Russia) and the United States, in what was to be viewed as the space race, where one nation sought to better the other and to be the first to achieve a major '*space-mark*.' This competition, occurring in this period, also recognised to be the '*Cold-War*' era, was fuelled by the distrust for the other nation and the broader ideological and military rivalry between the two superpowers, who both sought to achieve superiority over the other. In this sense, outer space served as another battleground and one which figuratively and literally had no boundaries.

2.1 *The Impact and Legacy of the Second World War*

The relationship of outer space and warfare should not be underestimated. Very few years had passed since the end of the Second World War (WW2: 1939–1945) when the Bumper-WAC went into space. WW2 (the War) had shown, only too clearly, the death and destruction that could occur on Earth. In many ways, WW2 served to demonstrate too that, even when nations had defined boundaries, they could be breached and national laws and sovereignty disrespected, on such a global scale, through invasion and other hostile acts.

Ironically, resources, in particular metals, were to play a significant part in warfare, being used to build military weapons and modes of transport that were at the forefront of delivering destructive means (Fox, 2021, 2022). The War was also instrumental in terms of legacies, and, perhaps, none more so than the use of an atomic bomb against a nation, which showed the sheer destructive capability of a weapon that could be delivered through the air (Fox, 2014, 2015a, 2015b). Weapon development, stemming from the War, was also fundamental in leading to man's exploration into space. The Bumper-WAC was in fact a rocket, a legacy of the war, that consisted of a JPL WAC Corporal missile sitting over a German-made V-2 rocket. The V-2 having been developed by Wernher von Braun's team of German researchers, who surrendered their research to the United States at the end of the War (NASA, 2008).

However, in the midst of the War, on 12 June 1941, representatives from Great Britain, Canada, Australia, New Zealand, the Union of South Africa, plus the exiled governments of Belgium, Czechoslovakia, Greece, Luxembourg, the Netherlands, Norway, Poland, Yugoslavia, and of General de Gaulle of France, met at St. James' Palace, London, where they signed a declaration which stated, (in part),

> That the only true basis of enduring peace is the willing co-operation of free peoples in a world in which, relieved of the menace of aggression, all may enjoy economic and social security; and that it is their intention to work together, and with other free peoples, both in war and peace to this end.

This declaration was arguably to lay the foundations for a collective body of nations to work together to ensure a peaceful existence on Earth. As can be seen, no mention was made outside of *our* 'world.' And, as the War was coming to an end, in April 1945, delegates of 50 nations met in San Franciso, for a conference, that finally led to the drafting of the Charter of the United Nations (UN) which was later adopted on 26 June, 1945 (United Nations—Preparatory Years).

The Charter is an instrument of international law, and UN Member States are *technically* said to be bound by it. It aims to codify the major principles of international relations, from sovereign equality of states to the prohibition of the use of force in international relations.

The Charter identifies four '*Purposes*' (Article 1). Summarised, these are as follows:

1. To maintain international peace and security. This includes taking effective collective measures for the prevention and removal of threats to the peace, and for the suppression of acts of aggression or other breaches of the peace, applying the principles of justice and international law, including the settlement of international disputes or situations which might lead to a breach of the peace;
2. "*To develop friendly relations among nations based on respect for the principle of equal rights and … to take other appropriate measures to strengthen universal peace*"
3. "*To achieve international co-operation in solving international problems of an economic, social, cultural, or humanitarian character, and in promoting and encouraging respect for human rights and for fundamental freedoms*" ……. and
4. To be a centre for harmonising the actions of nations in the attainment of these common goals.

To this end, there is a degree of overlap with outer space, given that activities in space initiate from Earth and have some form of oversight from Earth. Although the Charter makes no specific reference to activities away from the Earth, it does identify that the United Nations—as the 'Organization'—makes recommendations for the co-ordination of the policies and activities of the '*specialised agencies*' (Article 58), while Article 57 refers to these in terms of their 'related fields.' This said, it was a number of years before space in the form of the United Nations of Outers Space Affairs (UNOOSA) was established as a specialised agency. The UN Charter, though, remains the overarching treaty above others and establishes a mechanism for dispute settlement (in particular see Articles 33, 36–38), wherein it is identified that disputes that are likely to endanger the maintenance of international peace and security should first be addressed through negotiation, mediation or other peaceful means. This therefore can be identified as having impact to disputes that might also concern space.

3 United Nations and Outer Space Affairs

It was not until 1958 that the UN General Assembly, through resolution 1348 (XIII), established an ad hoc Committee on the Peaceful Uses of Outer Space (COPUOS) to consider the activities and resources of the United Nations in this regard. This was prompted by several key events in 1957 and 1958. In 1957 the Soviet Union launched the first artificial satellite, Sputnik, into space (4 October), and a month later Sputnik 2 launched with a dog named Laika on board. Then, on 31 January 1958, Explorer 1 was launched by the United States (ALIC). At this time there were no international laws specifically relating to outer space and the conduct by which space exploration and other activities should be pursued. UNOOSA was also established in 1958 as a specialised agency to assist and advise (COPUOS) and other international bodies relating to the peaceful uses of outer space (General Assembly in its resolution 1348 (XIII) of 13 December 1958) (Fox, 2016, 2022, 2023).

It is questionable whether consideration should have been given earlier to the need for a dedicated mechanism to regulate outer space in terms of peace, security and other activities in space, and indeed, even whether the UN Treaty could have, or should have, been phrased differently to extend beyond the remit of our world. While the *Treaty on Principles Governing the Activities of States in the Exploration and Use of Outer Space, including the Moon and Other Celestial Bodies*' (OST) transpired in 1967, the fear of nuclear war in space had been potentially realised some years earlier when a number of nations had tested and, therefore, detonated weapons, which not only had an impact on Earth but also on space (Fox, 2023).

3.1 Roles and Responsibilities

UNOOSA, today, identifies that it helps all countries, especially developing countries, access and leverage the benefits of space to accelerate sustainable development. While working towards this goal through a variety of activities that cover all aspects related to space. This includes space law through to space applications. Linking to the wider objectives of the UN Charter, UNOOSA both manages and implements the programme on the peaceful uses of outer space. This being aimed at strengthening international cooperation in space activities and in the use of space science and technology for achieving sustainable development (UNOOSA). An interconnected role relates to aiding nations increase their capacity to draft or revise national space law and policy in line with international normative frameworks on space. UNOOSA stresses that this is particularly important as more and more *actors* enter the space arena. However, here there are several key factors to bear in mind. Firstly, the UN Charter, and hence the related governance of space by subsequent treaties and agreements, applies to Member nations or States (the signatories and ratifiers) rather than private bodies—so it is questionable whether there is a limit in terms of the applicability to all '*new actors*.' Secondly, the UN Charter codifies the major principles of international relations (UN Charter) while, more widely, codification is generally understood as the more precise formulation and systematisation of rules of international law on subjects that have already been extensively covered by State practice, precedent and doctrine (Milde, 2012). In this regard, it would be noticed that the UN OST came before most countries had any national laws relating to space and therefore did not codify existing laws or even practices. This said, the United States had, nearly 10 years earlier, passed a national space law in the form of the National Aeronautics and Space Act of 1958, which potentially could be viewed as leading developments at a UN and international level. Certainly, it might aid to account for the viewpoint attributed to the United States, which at times, has been identified as differing from the OST or the perceived spirit it intended (commented further below).

4 Outer Space: The Relevance to Resources and Mining

The world has noticeably changed since the 1950's. This has implications to space too. While ensuring peace on Earth remains a constant factor, this is, more than ever, impacted upon by technology, including that placed in outer space. The symbiotic linkage of space to war, and vice versa, remains only to clear, While the wider synergetic relationship to mining might not be as apparent or transparent, it is argued that today having access to minerals is key to exerting superiority over another and that includes the access to resources that lie in space (Fox, 2016, 2022). Recent warfare on Earth (such as the Russia-Ukraine conflict) and the posturing of the U.S. President Trump to acquire resource rich nations and territories (such as Greenland and Canada), exemplifies the fact that States, including the original Cold-War nations, still have the determination to be seen as a wealthy superpower and take actions that further this goal. And hence, mineral wealth truly is the superpower of today (Fox, 2022).

Coupled with this, the superpowers still have the determination to be the first to achieve pursuits in space. On 20 January 2025, during President Trump's second inaugural address, he proclaimed that we (the U.S.) "*will pursue our manifest destiny into the stars, launching American astronauts to plant the Stars and Stripes on the planet Mars.*" This bold ambition chose to clearly replicate the 1969 Moon landing, whereby the U.S. firmly planted their national flag. This action has been questioned in terms of meaning, including any associated rights and claims, including to the resources and mineral wealth that lie on and in the Moon (Fox, 2016, 2022).

Debatably, the significance of mining in space was not fully appreciated or even predicted in the 1950s and 1960s, and therefore, guidance could be viewed as lacking or being imprecise in international agreements. Certainly, the involvement of non-State actors was not anticipated and the significance of sustainability to both Earth and space was not envisaged. Sustainability is now also increasingly spreading into the realms of space.

4.1 International Space Law: The OST

Alongside the UN OST and the later key treaties, there have been other declarations and principles, with some being aimed at specific sectors or use, for example, the Broadcasting Principles (1982) and the Nuclear Power Sources in Outer Space (1992). However, there have been no similar approaches taken regarding mining in space. Hence, from an international perspective, at least, the OST remains the main source to consider in terms of the legitimacy and acceptance of mining in space.

Firstly, it should be noted that the OST fails to specifically mention mining or other commercial activities relating, more broadly, to resource extraction. However, it affirms the fact that outer space (and, hence, that contained within) is the common heritage of humankind. More specifically, it is stated in terms that there is a '*common interest of all mankind in the progress of the exploration and use of outer space for peaceful purposes*' (Preamble). Exploration is further addressed in the first Article, which makes specific reference to the Moon and other celestial bodies. This being essential in terms of emphasising that it is to the OST that guidance must be sought and that the Moon Treaty contains only limited expansion in relation to the Moon. This said, reinforcement must also be given to the differences in terms of acceptance of the Moon Treaty, particularly in comparison to the OST. As of January 2025, there were 116 nations which had ratified the OST, while there were another 23 signatories to the Treaty. This compares with only 17 nations ratifying the Moon Treaty and 4 nations signing the Treaty (UN Status of Agreement document).

The OST yet contains only seventeen articles, which makes it a slim document and wherein many of the articles are subject to various interpretations, in terms of not only the meaning but the various levels of acceptance, or degree, for the '*permissible*' related activities.

The word '*exploration*,' as per the Treaty title, is used several times, as is the word '*use,*' and, while exploration may be '*free,*' it is also identified that this is for States to undertake. Of course, in this regard, it could be questioned whether this approach is only applicable to those that have ratified the Treaty. For instance: what is the impact on nations that have only signed the OST? And, what of the nations that have (to date) neither signed nor ratified the OST? Would their right or entitlement be any different? Particularly given the fact that the OST identifies that outer space, celestial bodies and the Moon should be accessible (free for exploration) '*without discrimination of any kind, on a basis of equality*' (Article I). So presumably this would also include those nations that may, in time, become space faring and, yet, have neither ratified nor signed the Treaty. This also leads to the questioning of corporations—such as multi-global entities and even national enterprises, and where they fit in terms of obligations of use/exploration of outer space, and, more significantly, their liability or the applicability of a State or States' liability. Again, this could vary depending upon nationality, citizenship or, more likely, formulation of a company (location-based) and/or even the launching State (points further commented upon below). Hence, it is multifaceted in terms of the position, status and ability of *new actors.*

There are further questions in terms of what *exploration,* of course, actually means, or perhaps more precisely, what it allows to be undertaken in this pursuit.

Applying a dictionary definition, the following could be understood; namely that exploration is:

> The activity of travelling to and around a place, especially one where you have never been or few people have been before in order to find out more about it (Cambridge Dictionary online).

And yet, the same Dictionary also identifies that an activity might be through '*an economic*' or '*business*' activity.

The Collins Dictionary online offers that exploration activities might include ones such as '*core testing and the drilling of test wells.*'

While the Oxford dictionary online offers that exploration may involve actions relating to gold, it is not fully specific as to what this might be.

Hence, the application of various dictionary interpretations epitomises the difficulties in terms of agreeing permissible activities to be undertaken in outer space, particularly also given that there are no negative statements in terms of forbidden activities within the OST. Even regarding nuclear weapons (and other weapons of mass destruction) the Treaty refers to the need to merely '*refrain*' from placing these in orbit around the Earth, or on a celestial body. Thus, the Treaty does not state that extraction, mining or other commercial activities are not permitted. However, this raises questions as to whether the intention is able to be deduced elsewhere within the Treaty, that is, as to whether '*taking something*' (regardless of it being extracted or mined) is acceptable. And, whether such an activity becomes even more contentious, or *frowned upon* if it is taken for a profit, that is, as part of a commercial activity.

In many ways, the OST is written from the stance of appealing to the good nature of States to keep the peace in outer space. This is somewhat ironic, given the legacy of space development to warfare and the inability to keep peace on Earth. It reinforces principles, while being a treaty, relating to a shared and common area—but without specifically saying this either. And, therein, lies some of the main issues in terms of an intention to create a '*Utopian Space*' but without clearly articulating how this is to be achieved.

From the UN perspective, outer space is intrinsically viewed as a *global common*, in the same way as the high seas, the atmosphere (the airspace over the high seas) and Antarctica are (Global Governance, 2015). And yet each of these three *global commons*, attributed to our Earth, have not been without conflict and disputes. The fundamental principle of a global common is that such a defined area presents a global resource domain, wherein there are common-pool resources which are not subject to national jurisdictions. However, in the absence of an authoritative definition, varied interpretations of the term global common also lead to an unclear meaning and, thus, inconsistent applications.... *and, hence, disagreements*. While the Treaty actually does not use the wording '*global common*' at all within.

Alongside specific reference to '*the common interest for all mankind*' rationale, the Preamble, however, does reaffirm the belief that '*exploration and use of outer space should be carried on for the benefit of all peoples irrespective of the degree of their economic or scientific development.*' Here again, the reference is not specific as to where the economic or scientific development should have occurred (on Earth or in outer space), but it could, however, lead to the interpretation, as per the Cambridge Dictionary (above), that *exploration and use* might indeed include economic activities. And, if this is the case, there is a clear indication that such associated advantage, or wealth, 'should' 'benefit' all. Again, wider reading might lead to further questions in terms of a somewhat 'ideal' and 'moral' approach which is only advocated *should* rather than be a 'must.' The Preamble also does not confirm where the 'peoples' are to be located; however, Article I does add some clarity, in both regards, whereby it is stated that both exploration and use of outer space (including the Moon and celestial bodies) '*shall be carried out for the benefit and in the interests of all countries....*' While also, again, referring to the aspects of both the degree of economic or scientific development, it is stated that '*the province*' is for all mankind to use. Hence, potentially two interpretations can arise:

(1) That regardless of economic and scientific developments on Earth, and, hence, the wealth of a nation, including to venture into space, there is an obligation that all nations have the opportunity to use and explore… and to 'benefit.' And/or,
(2) That regardless of economic and scientific developments in outer space, there is an obligation that all nations have the opportunity to use and explore, and to 'benefit.'

'Benefit' perhaps implying that nations that acquire wealth (whether monetary—economic; or scientific) due to space use/exploration, are obligated to—'*shall*'—share.

Article II, however, adds another layer of complexity to discussions and hence varying interpretations as to what the OST intends, whereby it is stated that '*Outer space, including the Moon and other celestial bodies, is not subject to national appropriation by claim of sovereignty, by means of use or occupation, or by any other means.*'

Once more, there are a number of aspects that can be questioned in this sentence as to the intended meaning. The key word here, to start with, is 'appropriation' and what is specifically meant; and, then, secondly, who is it applied to?

Again, the most commonly applied definition for appropriation refers to '*the act of taking something for your own use….*' (Cambridge Dictionary online). While this is perhaps without differences in terms of a dictionary interpretation and appears a widely recognised principle, it nonetheless remains, once again, vague and subject to various legal factors and, thus, interpretations. For instance, what exactly is the '*something*' that cannot be used?

If, in this instance, hypothetical consideration is given to the 'something' being a resource (maybe, for example—acquired through mining)—and an appropriation of property definition from the Theft Act (UK-1968) is utilised, then there may be a need for clarity, or a redefinition in terms of what actually is the '*taken*' (or mined) substance/resource in the first place. For example, in the same way differences are recognised in terms of electricity (*a resource*) being abstracted. This would also lead to further questions being asked as to any potential wrongdoing if the 'use' remained in outer space—meaning the acquired or appropriated resource stays, or is, even, used there?

Hence there are a number of jurisdiction elements to also be considered, including how far and in what sense the resource (mineral or otherwise) is classed as being or having been 'taken?' Who would have control, or say, over 'perceived' wrong-doings? And thus, going full circle—ultimately, questions would again arise as to whether there would essentially be any wrong-doing in the first instance.

An extended dictionary (and legal-UK) definition moreover refers to any appropriation, *wrong-doing* necessitating that the '*something*' actually needs to belong to someone else in the first place (Oxford Learner's Dictionaries) and that the '*something*' is usually taken without permission (Vocabulary.com).

In terms of the acquired space resource, it would have to be questioned who owned it before it was 'appropriated?' Given that the province is for all 'mankind,' it could even be argued that there would already be an entitlement, and, hence, no offence or *wrongdoing* would actually occur. Thus, a simple statement (as per Article II) from an Earth context becomes complicated when applied to an outer space domain.

In this instance, Article II further states that the appropriation cannot be a national one, more specifically referring to a claim by a sovereign nation. Though this has led to various interpretations and, hence, much debate again transpiring. Firstly, it is viewed that, as Article II is silent as to other actors and private entities, only addressing sovereign nations, there would be no prohibition of private appropriation of outer space resources (Wasser & Jobes, 2008). Conversely though, it could also be deduced that private entities would have no entitlement, as the OST simply excludes them. While a wider interpretation is that Article II governs all actors, private and State, thus prohibiting all forms of appropriation in outer space (Cherian & Abraham, 2007; Tronchetti, 2015). This is certainly the view of Judge Lachs, a former Chairman of the Legal Subcommittee of the COPOUS, who identified that Article II does not only refer to a prohibition on *national appropriation* but includes both "sovereign rights and private property rights" (Goedhius, 1970; Jakhu, 2006). And, presumably, private property rights and, hence, appropriation would extend beyond not just a nation but a multi-global entity which appropriated. However, this ultimately remains undetermined.

In effect, Article II reinforces that outer space is established as a supranational area domain, that is beyond national territorial sovereignty (Jakhu & Freeland, 2022), whereby appropriation cannot serve the exclusive interests of any State (or even State entity), but it is still argued as to the implementation of the OST, and particularly Article II, to a global corporation. Principally, given that Article VI identifies that '*States Parties to the Treaty shall bear international responsibility for national activities in outer space, including the Moon and other celestial bodies, whether such activities are carried on by government agencies or by non-governmental entities*...' however, here emphasis is clearly accorded to *national* activities and furthermore identifies that the activities '*shall require*' both authorisation and continuing supervision by the appropriate State Party to the Treaty. In term of '*international organisations*' carrying out activities, it is stated that responsibility for compliance with the OST shall be the responsibility of the international organisation and by the States Parties to the Treaty that are participating in such events. However, here there is a possible lacunae in terms of international global corporations and whether these are factored in as an international organisation—for instance, was the intention to focus on a body, such as the European Space Agency (ESA), which sees nations (space agencies) coming together to collaborate? Or, did it foresee the likelihood of a more private global entity entering into the space arena? And what of any formation of global conglomerations that in the future consists of nations who have not ratified the Treaty?

Article IV, of the OST, refers specifically to the Moon (and other celestial bodies) and identifies that the '*use of any equipment or facility necessary for peaceful exploration ...... shall ... not be prohibited*.' In many ways, this seems to conflict, in parts that is, with Article II, whereby the latter article now says it is permissible to both use equipment and place a 'facility', *however defined*, onto the Moon. Again, the full meaning of exploration is not readily understood or agreed, and thus it would also lead to the re-questioning, as within Article II, as to what is able (*if anything*) to be appropriated; plus, what the difference is in terms of a 'facility' being allowed, but not its use or occupancy?

While the OST has received significantly more agreement, and hence, endorsements (that is, in the form of both signatories and ratifications) than the MA, for instance, it remains far from being clear and transparent. This has led to varying interpretations and opinions, by both nations and other actors, being expressed. In many instances, certainly those with a spacefaring capability and capacity, a more *open access* attitude seems to prevail.

5 The Moon Agreement

However, trying to address some of the grey areas in the OST, in the latter part of 1979, the United Nations drafted and approved a treaty that would establish guidelines on future uses of Earth's moon and outer space. The full title is the "Agreement Governing the Activities of States on the Moon and Other Celestial Bodies." It was adopted by the U.N. General Assembly without opposition, though the Moon Agreement (MA) was certainly not welcomed by States in the same way as the OST and remains equally unpopular today. Ironically, this is predominantly due to the fact that it tried to put meat back onto the bone, that is the bare framework of the OST, and provide more details than exist within the OST.

While in the United States, the then President Carter administration supported approving the MA Treaty, the election of President Reagan scuppered any future advancement of it, with it receiving heavy opposition from business and scientific communities alike. This was largely due to critics identifying that it would prohibit commercial development in outer space and would "socialize" future lunar and planetary bases and space stations (NSS, 1982).

Article 11 of the MA, in particular, added clarification in terms of natural resources, stating that these are the common heritage of mankind. It went further to identify that these are not to become property and clarified this with respect to both States and other entities, whereby it stated that appropriation meant to '*any State, international intergovernmental or non-governmental organisation, national organisation or non-governmental entity or any natural person*' (Article 11.3).

While the MA did not seek to prohibit the collection, or even mining of resources, it did convey that any activity must be compatible with the Treaty, identifying in particular Article 11, paragraph 7, which concerns:

(a) "The orderly and safe development of the natural resources of the Moon;
(b) The rational management of these resources;
(c) The expansion of opportunities in the use of these resources;"

And perhaps, most controversial of all of these was the fourth requirement, which referred to

(d) *"[An] equitable sharing by all State Parties in the benefits derived from those resources, whereby the interests and needs of the developing countries as well as the efforts of those countries which have contributed either directly or indirectly to the exploration of the Moon, shall be given special consideration."*

Like the wording within the OST, the MA did not specifically seek to enforce this; moreover, it really spoke of an ideal, advocating *or appealing* to nations to give serious and '*special*' consideration to this as a principle. The industry critics, on the other hand, particularly in the United States, deemed this to be *a giveaway of unprecedented proportions,* whereby the approach was viewed as a mechanism and control to stifle, if not prevent, private industry from developing outer space or certainly realising a profit by doing so (NSS, 1982). Whereas it could, perhaps, be viewed in terms of the converse argument, that relates to recognising that commercial activities, which had not to date been addressed, would run the potential for territorial disputes or even neo-colonial exploitation without intervention mechanisms being in place. Hence, this advocated action 'relating to equity sharing,' far from suppressing development, should be viewed as a humanitarian advancement which encourages collective pooling to develop, manage and share during respective steps.

In other words, the MA sought, in part, to democratise resource acquirement from the Moon and other celestial bodies, and, arguably, to go one step further towards creating a utopia in space. What it proposed was the establishment of an "international regulatory regime" to create and oversee an equitable system to share resources, once lunar exploitation could be realised. However, in reality, it ultimately lacked, and still does, the depth needed to articulate this and hence to achieve this; or, even to gain the consensus to agree to this approach being further explored.

The language of the MA has clear parallels to the United Nations Convention on the Law of the Sea Treaty (UNCLOS), which at the time was neither settled nor agreed, with the United States being particularly vocal over comparable concerns in terms of its use and entitlements (Fox, 2016, 2022). In a similar manner to the MA, UNCLOS sets out the legal framework for the seas and the oceans by defining the rights and obligations of States Parties with respect to the maritime environment. Hence, the main function is the advancement of the peaceful use of the seas, while regulating the use of marine resources and promoting the conservation of living resources together with the preservation of the marine environment. While the Convention went on to be ratified by 170 parties (to this day—May 2025), the United States has neither signed nor ratified it. The major factors for not doing so relate to objections to UNCLOS' provisions concerning the seabed and ocean floor, including mining and wealth sharing, which it considers are against both the State's economic and security interests (Fox, 2022). This, in essence, mirrors the argument and dissatisfaction of the MA. Incongruously, the other two superpowers of today, Russia and China, have both signed and ratified UNCLOS, unlike the MA, which neither Russia, China nor the United States have either ratified or signed. The MA therefore remains the least successful international space treaty, with only 17 States being Party to it; none of these, however, are space-faring ones (Tronchetti, 2024).

Ironically, this is one State less than it achieved when it opened for signatures on 18 December 1979.

5.1 Withdrawing from International Agreements

Before 5 January 2023, no State had ever withdrawn from any of the five United Nations treaties on space-related activities. However, the unprecedented action taken by Saudi Arabia in terms of withdrawing from the MA typified the sentiments of States that had not seen fit to either sign of ratify it in the first instance. In many ways, this decisive action though could be viewed as more damming or, certainly, reflective of the way of thinking in terms of reducing or preventing controls over resources both on the Moon and elsewhere in space. The actual thinking of Saudia Arabia remains unknown; the nation is not required to provide its reasons for withdrawing from the Moon Agreement. Article 20 only stipulates that any State Party to the Agreement may give notice of its withdrawal, one year after its entry into force, by written notification to the secretary general of the United Nations. And, that such withdrawal can take effect one year from the date of receipt of this notification.

One of the most logical explanation, for the nation withdrawing, is that the Moon Agreement limits or impedes the nations own ambitions. And this reason, and perhaps action, will perhaps, become more relevant as nations who did not previously have a space presence, develop, or pursue R&D opportunities—including mining and accessing resources.

Similarities likewise remain with UNCLOS, whereby nation signatories have equally expressed dissatisfaction and therefore stated an intention to withdraw. In 2024 Russia was particularly vocal in stating this, and many of the associated reasons have parallels with the MA, although the MA did not perhaps state intentions and a governance system as clearly as UNCLOS, which endured considerable years of debate before being accepted (Salinas, 2024). UNCLOS eventually entered into force in 1995, although Russia took more than two years to ratify it, expressing doubts regarding Part XI of the Convention, which established a regime for the deep seabed, and, equally, its ability to protect the nation's interests (Cirkovic, 2018). This being largely centred around the concept of common heritage of mankind (Karev, 1995).

UNCLOS is based on customary international law, but was perhaps able to draw upon, to define and to codify standards and principles of international maritime law far easier than the MA. This is due to the more established historic legacies of the sea, and thus the ability of UNCLOS to establish a comprehensive regime that defines the governing rules relating to the sea, its use and resources (IMO; Fox, 2022). It establishes, therefore, a framework for managing relations, developments and disputes, including a dedicated mechanism in the form of the International Tribunal for the Law of the Sea (ITOS), the International Seabed Authority (ISA) and the Commission on the Limits of the Continental Shelf (CLCS). This said, disputes continue, for example, relating to historical rights, as is the case in the South China Sea, and also wider aspects relating to national security, which ultimately stands to call into question the Conventions effectiveness (Bateman, 2016; Fox, 2022). Alongside the impact of legacy factors, new disputes also arise, largely relating to the deep seabed and, ultimately, the access to the resource wealth that lies below (Fox, 2022; Østhagen, 2020).

Like the MA, UNCLOS has a clause that allows States to leave, as per Article 317, although no party has yet made use of it (Salinas, 2024). However, the withdrawal from Saudia Arabia to the MA could yet see other nations not only leaving this Agreement but UNCLOS and the OST, which also replicates this entitlement within Article XVI.

It should be recalled that when the OST and the MA were approved the realities of mining space resources was not feasible; and, yet today, the probability of one day doing so is now envisaged and is likely a reality, in the not-too-distant future. However, lessons learnt from Earth indicate that disputes will certainly arise and that no mechanism will be agreed beforehand that ideally prevents conflict, or is equally specific in terms of resolving mining, acquiring and sharing disputes. Even if, by some miracle, a system was put in place there unequivocally remains the potential for later expressing dissatisfaction and even withdrawing from a promise to commit (Fox, 2016, 2022).

6 Expanding Beyond International Law

When the UN wrote the OST few nations had ventured into space and space activities were by no means to the level they are today. Many other nations have since also become space faring, while others have utilised other nations modes and means to have their own presence in space—such as satellites. Private entities that have also entered the mix, and space law has dramatically increased. While the UNOOSA (as per Sect. 3.1. of this chapter) actively aids all countries, especially developing countries, access and leverage the benefits of space to accelerate sustainable development, there ultimately remains some degree of conflict in terms of ensuring that space development does not compromise sustainability not only to the Earth but also other areas within space. Responsible use of space remains key, and this includes ensuring that both relevant space law and policies are in place at an international and national level. The irony is that without such agreements, peace is ultimately in jeopardy both in space and on Earth.

6.1 *Resources:* Politics and Conflict.....

The Earth continues to revolve in space, and it continues to experience disagreement and warfare. Many of the reasons for needing international space law treaties and agreements apply equally today as they did in the 1950's, 1960's and 1970's; and, it remains questionable whether we are, as such, in another Cold-War period

(Fox, 2024b). Nations remain divided and annexed by various different agreements. Certainly, the Earth remains on a circular journey wherein history frequently repeats, particularly from a negative perspective. Colonial wars and land claims were often driven by the relentless pursuit of resources. Stemming back centuries, from the ancient conquests of land to trade routes, geopolitical (resource) struggles have continued, whereby they have, more recently, been centred around access to natural gas, oil, and other mineral wealth. Michael Klare, at the turn of the century, argued that competition for natural resources remains a primary cause of *global* conflict, yet, while the control over resources has and will continue to shape global politics and foreign policy dimensions on Earth, it will also increasingly be influenced by space and principally the wealth that lies above us (Fox, 2016, 2022). This remains particularly so due to increasing resource scarcity on our planet. Ironically, the drive for a sustainable future increasingly necessitates policies that utilise rarer resources that either lie at the bottom of the ocean, or above us, in space (Fox, 2016, 2022). A natural and inevitable deduction remains that States will seek to 'acquire' such wealth in order to remain sustainably dominant.

Achieving consensus relating to the MA seems a lost course, save from a negative perspective in terms of it being accepted that no further nations will ratify it. While achieving any new international space law seems an impossibility. This therefore calls into question where international law now fits in. Even though nations cannot readily agree to an international approach, States are though increasingly seeing the need for their own laws that actively seek to govern space. However, it remains questionable whether this actually aids or, potentially, adds a further layer of complexity, particularly when viewed from a mining and resources perspective. Although the UNOOSA identifies that it assists countries to understand the fundamentals of international space law, the fact that the OST has been interpreted in many different ways by States speaks to the fact that it is invariably viewed from a stance that benefits a nation rather than a collective and unified world. And, while the UNOOSA further states that it will help nations to '*increase their capacity to draft or revise national space law and policy, in line with international normative frameworks on space*', there is no obligation on States to approach the UNOOSA and seek such assistance, or, if they do, to follow any guidance provided. This has led the UNOOSA to recognise and accept that States have enacted national space legislations that have taken a number of different approaches in dealing with national space activities. While, furthermore, some States have adopted their own national legal frameworks according to '*the specific needs and practical considerations of the range of space activities conducted and the level of involvement of non-governmental entities.*' This increasingly sees commercial activities being catered for, while some nations have specifically enacted space legislation and laws relating to resources in space. Increasingly, from a national perspective, this entails more private party involvement and reference to private property rights over resources extracted from space. The United States, in 2015, was the first nation to have taken this direction, with the signing of the US Commercial Space Launch Competitiveness Act. Yet even national laws have been the subject of various interpretations within the nation and external to it—that is by other nations. Although the United States has now gone one step further seeking to questionably promote international cooperation instead of conflict regarding activities in space, through the 2020 Artemis Accords. While the Accords are, however, a non-binding set of principles designed to facilitate civil space exploration, and, arguably, exploitation, they again remain highly controversial in terms of not only the content and intent, but more specifically the division of nations—those that have signed or, more expressly, been invited to sign and those that have not (Fox, 2016, 2022)—(this is further discussed in Chapter 6 of this publication: Fox).

7 Conclusion

This chapter set out to explore the United Nations Outer Space Treaty and hence, the relevance to space mining. It questioned to what degree space mining was potentially overtly or covertly covered, and/or to the possible guidance or interpretation that could be formed or deduced from the contents of the OST. While there is no doubting that the OST remains the international Magna Carta for space, it fails to go into the depths needed, establishing only a basic approach concerning activities in outer space. Many provisions of the Treaty have variable interpretations, which are often interpreted by nations to their benefit. This is becoming increasingly so, as the possibility of extracting resources from space, together with the incentive to do so, becomes more elevated in terms of national priorities.

It was arguably a failing in the 1960's to develop the OST in terms of providing clarification and interpretation, and there was certainly a lack of foresight to realise the potential for Earth hostilities to escalate into space. This is perhaps particularly so now given the realisation of the wealth of resources that are in outer space.

As was discussed, history does have a tendency to repeat itself, and this arguably includes the resurgence of tensions on Earth, particularly over the last few years. This has noticeably been heightened since the Russian conflict with Ukraine and what is witnessed today is reminiscent of the Cold-War period, wherein nations are increasingly taking sides, and whereby the two dominant ones are the United States on one side and Russia on the other.

At the same time, there is arguably a new space race, which centres around commercialisation of activities. Increasing becoming central to this is the possibility, or more probable the likelihood, of nations seeking to extract resources from space. Space remains rich in terms of various resources, many of which are scarce or even do not exist on Earth. Certainly, resources are not found equally distributed across the globe and hence nations with the most needed and desirable resources are likely to become the superpowers, the wealthy, dominant countries. In a bid to counteract this, and to try to ensure a more balanced world, those nations with technological know-how, that have the capability and capacity, are likely to look elsewhere. Given the historic legacy on Earth, this will almost certainly lead to conflict, including land (ownership) claims and even invasions. Alongside this, global commons and areas deemed to be humankind's heritage will be challenged—this includes the ocean seabed and Antarctica. Ironically, as conflicts develop and particularly whereby they remain unresolved, acts of hostility and warfare will likely occur which will only increase the need for more access to resources—including minerals and metals, in order to defend or maintain national security. Space therefore remains another domain to potentially defend or for warfare to escalate into.

Nations clearly have a thirst to venture into space, while space also remains a target for resources, including wealth acquisition, whether in peace or in times of conflict. Initially, when this occurs, the resources are more than likely to remain in space and even to be utilised there, perhaps in support of the pursuit to travel to Mars. This, however, is also likely to be positioned in terms of a space race and competition between nations. The legalities and even consensus of actions in space, particularly of a commercial nature, remain unclear and none more so than where nations (individual or collective) and/or not involving private entities, acquire and use resources obtained from and in outer space. There remains a myriad of scenarios—some of which have been identified and interpreted in terms of the application of various definitions, relating to *acquiring* and/or *using* resources. This has also included in terms of the implications and challenges *of* and *over* jurisdiction.

Alongside questions as to whether, and in what circumstances, acquiring space resources is or would be permissible, from an international stance, there is also the extended consideration in terms of benefit sharing. The reality, however, is that achieving consensus on this, and, more importantly, actually undertaking any form of sharing wealth acquired from space, is likely never to be achieved. Existing international agreements remain debated among nations and none more so than those that try to create an ideal or utopia, where equity, particularly relating to tangible wealth, can be distributed. The reality is that international space law is simply not equipped to deal with space resources including space mining.

Given this, and given the lack of international agreement in terms of permitted actions relating to space resources and an appropriate governance system, nations are beginning to formulate their own approaches to not only space exploration but exploitation. However, there are now inconsistencies in many of the national approaches, which also complicates the ability to ever achieve harmonisation and agreement.

In many ways, nations are testing the boundaries of not only international treaties but outer space too.

References

ALIC - Archives Library Information Center (U.S.) – Space Exploration. https://www.archives.gov/research/alic/reference/space-timeline.html

Bateman, S. (2016). UNCLOS and its Limitations as the Foundation for a Regional Maritime Security Regime. https://dr.ntu.edu.sg/entities/publication/dd308d41-4135-45fc-b8ed-5d1e311aebad

Cambridge Dictionary online. https://dictionary.cambridge.org/dictionary/english/exploration, https://dictionary.cambridge.org/dictionary/english/appropriation

Cherian, J. G., & Abraham, J. (2007). Concept of private property in space – an analysis. *Journal of International Commercial Law and Technology, 2,* 211, 213.
Cirkovic, E. (2018). Russia, international law, and the melting of the Arctic, Völkerrechtsblog: international law & international legal thought. https://doi.org/10.17176/20180113-18262
Collins Dictionary online. https://www.collinsdictionary.com/dictionary/english/exploration
Commission on the Limits of the Continental Shelf (CLCS). https://www.un.org/depts/los/clcs_new/clcs_home.htm
Finch, E. R. (1984). Magna Charta of outer space for all nations. *Acta Astronautica, 11*(6), 337.
Fox, S. J. (2014). The evolution of aviation: In times of war and peace – blood tears and salvation!*International Journal on World Peace, XXXI*(4), 49–79.
Fox, S. J. (2015a). 'CONTEST'ing Chicago. origins and reflections: Lest we forget!*International Journal of Private Law, 8*(1), 73–98.
Fox, S. J. (2015b). To practice justice and right' – international aviation liability: Have lessons been learnt? *International Journal of Public Law and Policy, 5*(2), 162–182. https://doi.org/10.1504/1JPLAP.2015.071027
Fox, S. J. (2016). SPACE: The race for mineral rights. 'The sky is no longer the limit.' Lessons from Earth. *Resources Policy, 49,* 165–178.
Fox, S. J. (2021). The nexus between resources and criminal activities: 'Recycling Crimes' (Metals). *Resources Policy, 74.* https://doi.org/10.1016/j.resourpol.2021.102377
Fox, S. J. (2022). Exploiting – land, sea and space: Mineral superpower' in the name of peace: A critical race to protect the depths and heights. *Resources Policy*, 79.
Fox, S. J. (2023). The UK's '*appetite*' for space: An increased craving!*Journal of Air Law and Commerce, 88*(4).
Fox, S. J. (2024a). Space Tourism – Ensuring a safe and equitable framework lessons learnt from the U.S. and aviation. *Journal of Air Law and Commerce, 89*(3). https://doi.org/10.25172/jalc.89.3.4
Fox, S. J. (2024b). The evolving 'Space' in the EU: A circular journey. *Annals of Air and Space Law, 48.*
Goedhius, D. (1970). Legal aspects of the utilization of outer space. *Netherlands International Law Review, 17,* 25. https://www.cambridge.org/core/journals/netherlands-international-law-review/article/abs/legal-aspects-of-the-utilization-of-outer-space/78E1F91D5A5C6E976A179E7E292EBDFD
IMO: International Maritime Organization: United Nations Convention on the Law of the Sea. https://www.imo.org/en/OurWork/Legal/Pages/UnitedNationsConventionOnTheLawOfTheSea.aspx
International Seabed Authority (ISA). https://www.isa.org.jm/about-isa/
International Tribunal for the Law of the Sea (ITLOS). https://www.itlos.org/en/main/the-tribunal/the-tribunal/
Jakhu, R. S. (2006). Legal issues relating to the global public interest in outer space. *Journal of Space Law, 32.*
Jakhu, R. S., & Freeland, S. (2022). The relationship between the outer space treaty and customary international law, 59th IISL colloquium on the law of outer space 2016 – legal perspectives on space resources and off. *Earth Minute, 5.* https://www.elevenjournals.com/tijdschrift/iisl/2016/2%20Legal%20Perspectives%20on%20Space%20Resources%20and%20Off-Earth%20Mining/IISL_2016_059_002_003
Karev, S. (1995). The Russian federation and the UN conference on the law of the sea. *Proceedings of the Annual Meeting (American Society of International Law), 89,* 455–458. http://www.jstor.org/stable/25658962
Mckinsey & Company, Space: The$1.8 trillion opportunity for global economic growth, 8 April, 2024.
Klare, M. T. (2002). *Resource wars: The new land- scape of global conflict.* Henry Holt.
Milde, M. (2012). *Essential air and space law* (2nd ed.). Eleven International Publishing.

NASA. (2008). First human-made object to enter space. https://www.nasa.gov/image-article/first-human-made-object-enter-space/#:~:text=In%201949%2C%20the%20%22Bumper%2D,%2Dmade%20V%2D2%20rocket

NSS.org: L5 News: UN Moon Treaty Falling to US Opposition Groups. https://nss.org/l5-news-un-moon-treaty-falling-to-us-opposition-groups/#:~:text=This%20controversial%20%E2%80%9CMoon%20Treaty%E2%80%9D%20elicited,planetary%20bases%20and%20space%20stations

News: UN Moon Treaty Falling to US.

Østhagen, A. (2020). Maritime boundary disputes: What are they and why do they matter?, *Marine Policy*, *120*, ISSN 0308-597X, https://doi.org/10.1016/j.marpol.2020.104118

Oxford Dictionary online. https://www.oed.com/dictionary/exploration_n

Oxford Learner's Dictionary (online). https://www.oxfordlearnersdictionaries.com/definition/english/appropriation

President Trump's Inaugural Address. 20 January, 2025.

Qizhi, H. (1997). The outer space treaty in perspective. *Journal of Space Law*, *25*, 93.

Salina, B. P. (2024). Navigating uncharted waters: Russia's potential exit from The Convention on the Law of the Sea. Finabel. https://finabel.org/wp-content/uploads/2024/08/Belens-IF.pdf

Troncgetti, F. (2015). Legal aspects of space resource utilization,. In F. von der Dunk (Ed.),*Handbook of space law* (vol. 769, pp. 779–81).

Tronchetti, F. (2024). Moon agreement: Key provisions. In B. Sandeepa Bhat, D. Ukey, & A. Variath (Eds.), *International space law in the new space era: Principles and challenges* (Oxford, 2024; online edn, Oxford Academic), https://doi.org/10.1093/9780198909415.003.0009

United Nations Preparatory Years. https://www.un.org/en/about-us/history-of-the-un/preparatory-years

UN Charter. https://www.un.org/en/about-us/un-charter. Status of International Agreements relating to activities in outer space as of 1 January, 2025. AC105_C2_2025_CRP09E.pdf Published 5 May, 2025.

UN-UNCLOS Treaty Page. https://treaties.un.org/pages/ViewDetailsIII.aspx?src=TREATY&mtdsg_no=XXI-6&chapter=21&Temp=mtdsg3&clang=_en#1

United States Commercial Space Launch Competitiveness Act. (2015). 51 U.S.C. §§ 51301–51303

Vocabulary.Com. https://www.vocabulary.com/dictionary/appropriation

Wasser, A., & Jobes, D. (2008). space settlements, property rights, and international law: Could a lunar settlement claim the lunar real estate it needs to survive. *Journal of Air Law and Commerce*, *73*, 44–45.

National Approaches to Space—*Resources & Mining!*

Sarah Jane Fox

Abstract The United States (U.S.) has often shown itself as a nation to be forward-thinking in terms of advancing space activities, and this includes the pursuit of space resources. It has been competitive in staying ahead of other nations and maintaining a leading role in outer space. This chapter sets out to review and explore the impact, direction, and influence of the U.S. in the domain of space activities, which precede a United Nations and international approach. This is undertaken through the discipline of law, policy, strategy, and, arguably, includes space politics or *astro-politics*. It finds that the U.S.'s approach, centred around space resource, has largely been driven by a lack of international agreement in terms of permissible actions tolerated, or prohibited, in outer space. Yet, the U.S. approach is not without criticism in terms of the legalities of extending domestic legislation into the space domain and, invariably, into a global common for humanity. That said, perhaps in a bid to counter some of these arguments, the U.S.'s more recent direction has been to bring in other players, or nations, that agree to its wider space strategy and policy movements – in the form of the Artemis Program. Hence the Artemis Accords is further discussed, before attention is turned to other nations' domestic legislation which adopts a similar stance to the U.S. Invariably, it is concluded that challenges are likely to occur to these approaches when space resource extraction or mining in space actually becomes a reality and occurs.

Keywords United States · Artemis Program/Accords · Ownership · NASA · Luxembourg · United Arab Emirates · Japan

---

S. J. Fox (✉)
Leicester Law School | Director for the Institute of Digital Culture, University of Leicester, Leicester, UK
e-mail: sarah.fox@leicester.ac.uk

H. Jahankhani et al. (eds.), *Space Mining*, Space Law and Policy,
https://doi.org/10.1007/978-981-95-5543-7_6

1 Introduction

The initial governance, or, more accurately stated, perhaps, 'guidance' for the '*do's*' and '*don'ts*' in space, was the Treaty on Principles Governing the Activities of States in the Exploration and Use of Outer Space, Including the Moon and Other Celestial Bodies (the Outer Space Treaty—OST), which entered into force in 1967.

At that time very few States had existing national legislation that covered activities in space. However, with the rapid advancement of technology, the space sector today presents growing economic opportunities with mounting interest from not only governments but research institutions and private companies and corporations. Involvement of private companies in space activities has not always been so welcomed or even encouraged as it is today. This growth, viewed as the '*New* Space Economy,' is pushing the boundaries of what is possible in space and how this is achieved (The Congress; ESA). One such activity includes mining and mineral/resource extractions and, with these ever-increasing opportunities, nations are pursuing their own legislative approach, one that ultimately benefits their own economy (Fox, 2023).

This chapter sets out to review and explore the development of national approaches which have lately specifically given emphasis to assuming ownership or rights over resources from space, with primary commentary and focus on the United States' (U.S.) impact, direction and influence. Discussion is given to the acceptance of these approaches and the possible conflict in terms of international treaties—particularly the OST (also see related discussions within this publication, in Chap. 5). This is explored from a legal, policy and strategic perspective, which invariably includes astro-politics.

2 *In the Beginning*: The U.S. Legacy

The U.S. was one of the early adopters of a national approach, in fact, with their intentions pre-dating any international treaties, by nearly ten years. In April 1958, President Dwight D. Eisenhower asked Congress to pass legislation to create a civilian space agency to oversee the nation's space activities.

2.1 The National Aeronautics and Space Act

The *National Aeronautics and Space Act* (NASA or the *Act*) was signed by the then President (Eisenhower) on 29 July 1958 and was an act to '*provide for research into problems of flight within and outside the earth's atmosphere, and for other purposes*' (NASA). It should be noted that the 'for other purposes' was not specifically defined but, like the later International Treaty, it spoke to the fact that '*activities in space should be devoted to peaceful purposes for the benefit of all mankind.*' In terms of a governance structure, it referred to the fact that such "*activities*" '*shall be the responsibility of, and shall be directed by, a civilian agency exercising control over aeronautical and space activities sponsored by the United States.*' Yet it was not specific in identifying or detailing what activities were and/or included, or the mechanism to be put in place to ensure that such activities (however defined) should be for the benefit of all man (human) kind. That said, the Act did provide a definitions section (103) which grouped activities across three broad undertakings; however—these related to '(A) research into, and the solution of, problems of flight within and outside the earth's atmosphere, (B) the development, construction, testing, and operation for research purposes of aeronautical and space vehicles,' and a potential catch-all group, namely via (C) which related to 'such other activities as may be required for the exploration of space.'

It also provided a holistic list, so as to *"contribute materially to one or more of the following objectives:*

1. The expansion of human knowledge of phenomena in the atmosphere and space;
2. The improvement of the usefulness, performance, speed, safety, and efficiency of aeronautical and space vehicles;
3. The development and operation of vehicles capable of carrying instruments, equipment, supplies, and living organisms through space;
4. The establishment of long-range studies of the potential benefits to be gained from, the opportunities for, and the problems involved in the utilisation of aeronautical and space activities for peaceful and scientific purposes;
5. The preservation of the role of the United States as a leader in aeronautical and space science and technology and in the application thereof to the conduct of peaceful activities within and outside the atmosphere;
6. The making available to agencies directly concerned with national defenses of discoveries that have military value or significance, and the furnishing by such agencies, to the civilian agency established to direct and control nonmilitary aeronautical and space activities, of information as to discoveries which have value or significance to that agency;
7. Cooperation by the United States with other nations and groups of nations in work done pursuant to this Act and in the peaceful application of the results thereof; and
8. The most effective utilisation of the scientific and engineering resources of the United States, with close cooperation among all interested agencies of the United States in order to avoid unnecessary duplication of effort, facilities, and equipment."

It is prudent to note the original language and, that while speaking of a more equitable approach, that across decades, society had evolved in terms of equality of the sexes. Today, such language would have, no doubt, seen more gender-neutral terminology being applied, that is, rather than mankind, for instances, to humankind. Yet, it is contended that the concept of benefitting humankind from a morale perspective has actually been eroded due to the realisation of the wealth that lies above us and that nations now seek to protect their own 'national' interests. This said, there was an early indication that (outer) space was a domain that nations felt a need to protect. The Act clearly identifying that while a civilian agency should have control over space activities, the underlying control would still be State *sponsored*. In other words, this really was indicative of the actions (Space Activities) being State led or even

State controlled, certainly, at the very least, it indicated that it was to be supported and financed by a State budget and/or mandate. While there was a further passage within the Act which referred to the role to be played by the Department of Defense, whereby it was stated, that:

> activities peculiar to or primarily associated with the development of weapons systems, military operations, or the defense of the United States (including the research and development necessary to make effective provision for the defense of the United States) shall be the responsibility of, and shall be directed by, the Department of Defense....

Further on, in the Act, emphasis was again accorded to this joint use of space for both civilian and defence purposes and indeed the overlaps whereby it is stated that discoveries that have military value or significance should be shared (point 6: Declaration of Policy and Definitions, within Title I).

Title II concerns the coordination of both aeronautical and space activities, whereby section 201(a) established the National Aeronautics and Space Council (hereinafter called the "Council") with the membership being stipulated in terms of the composition. As enacted, key roles were identified regarding:

(1) the President (who resided over meetings of the Council);
(2) the Secretary of State;
(3) the Secretary of Defence
(4) the Administrator of the National Aeronautics and Space Administration; and
(5) the Chairman of the Atomic Energy Commission; (while other persons, sectors also participated).

A significant part of the Act, and therefore the governance for space in the U.S., was the creation or establishment of the National Aeronautics and Space Administration (hereinafter called the "Administration" or more frequently referred to, or known, as—NASA). The Administrator from NASA sitting on the Council—as per (4) above. NASA opened for business on 1 October 1958, with the Act identifying how the Administration would carry out the activities as well as providing information relating to some of the funding mechanism, while it also identified who should be employed. This granular detail is again significant in terms of identifying the closeness of civilian activities with the military and defence—whereby it is said, that in order to carry out the purpose of the Act, the Administration '*shall employ retired commissioned officers of the armed forces of the United States....*' Reference is also made to a joint committee which is a 'Civilian-Military Liaison Committee' (C-MLC, Section 204). The difficulties in clearly separating the position of space and aviation activities are also echoed through the Act too, which furthermore emphasises the close relationship between both the civilian and military sector within the space and aviation domains. In this respect, it should also be noted that no clear separation has ever been agreed in terms of where airspace (and law) finishes and the (outer) space domain starts. In the section of the Act on the C-MLC, reference is made to the Administration and the Department of Defense working together through the Liaison Committee and their role, wherein it is said that they '*shall advise and consult with each other on all matters within their respective jurisdictions relating to aeronautical and space activities and shall keep each other fully and currently informed with respect to such activities.*' This synergy, and at times difficulty, is further echoed in terms of the role of the Federal Aviation Administration (FAA) and their overlap in terms of space and aviation pursuits (Fox, 2024).

While the Act makes reference to engagement external to the U.S., in terms of 'International Cooperation' (Section 205) this is rather limited in terms of dimensions and the possible activities envisaged. What is perhaps more telling is the oversight of the government in this regard, despite the earlier reference to a *civilian agency* (NASA) *exercising control over aeronautical and space activities,* it is clearly stated that the Administration (NASA) but *under the foreign policy guidance of the President*, may engage in a program of international cooperation in work done pursuant to the Act. Yet the Act remains silent in terms of many of the activities that this could be and it is unlikely, that when written, the full extent of possible activities was envisaged. What is does however stress is that this should be 'peaceful applications' which must be undertaken with the agreement of the President with advice and consent of the Senate. NASA, furthermore, is subject to oversight in terms of a reporting process to the U.S. Congress across the field of aeronautics and space activities.

Over the years, changes have been made to the Act. However, from the onset, the Act, perhaps to be expected given the era and Cold-War tensions, also contained a specific section relating to security, although, arguably, what perhaps is surprising is the reference to the Atomic Energy Commission and the realms of data sharing between the parties.

The Act was also written to contain a short section on Intellectual Property Right (IPR) but again the depth of application is not specific in this regard. That said, it provided that both employee inventions as well as private contractor innovations brought about through space travel would be subject to government ownership. As is referred to in the objectives (as objective 5) the clear intention is for the U.S. to be a leader in space, while this is perhaps alluded to from a peaceful perspective, there can remain little doubt that the U.S. intention includes benefitting nationally from the benefits that space pursuits bring, which would also extend to the realms of protectionism and protecting national interests.

Amendments, over time, have certainly reinforced the aspect of commercialising space, with it being inserted into the Act, that '*Congress declares that the general welfare of the United States requires that the National Aeronautics and Space Administration (.....) seek and encourage, to the maximum extent possible, the fullest commercial use of space.*'

The definition section (at 103) has also been slightly refined to include a new inserted point (at C) which refers to the operation of a space transportation system including the Space Shuttle, upper stages, space platforms and related equipment. In this respect, during the lifetime of the Space Shuttle, the government remained the exclusive provider of space transport, with the revisions to the Act effectively dejecting the development of space travel by the private sector.

While mining and mineral/resource extraction was not specifically referred to in the original Act (as enacted) or subsequent amendments, the potential for such activities certainly became anticipated. Again, this perhaps also emphasises the intention of the U.S. to be a key leader in space pursuits and to prosper from the wealth of space. This is further reinforced in terms of a ninth objective that was added in, which referred to '*[t]he preservation of the United States preeminent position in aeronautics and space through research and technology development related to associated manufacturing processes.*'

Like the FAA, whose role relating to space activities has been impacted upon by various Presidents, the space community has watched each presidential transition to see how the new administration would handle decision-making on space policy and strategy. One such noticeable change has been the role of the C-MLC, which was abolished. While respective Presidents have made their own stamp in terms of the National Aeronautics and Space Council and even a National Space Council (via *the Executive Office of the President*—EOP) (Vedda, 2016).

The Act has become noticeably longer overtime, incorporating new areas relating to insurance alongside malpractice and negligence claims. It also has been expanded to incorporate new titles in terms of research. Though, the original Act (as enacted) came before any international (United Nations) treaties and therefore could not conflict with any subsequent agreements as such. In many ways it could be viewed as impactful, or even globally leading, in terms of the content, albeit that it was sovereign focussed, protecting national interests. Yet, this Act was not the first U.S. Act to show foresight into the opportunities presented by Space.

2.2 The U.S. and Communications

In 1934, way before humankind had ventured into space, the U.S. passed The Communications Act of (1934), which combined and organised federal regulation of telephone, telegraph and radio communications. The Act created the Federal Communications Commission (FCC) to oversee and regulate these industries and overtime it has been updated to include provisions governing new communications technologies, which includes satellite enhanced services. Once more, this shows the significance and relevance to space activities and the link to commercial pursuits. While, in 1962, again before the International (UN) OST, the U.S. wrote into their national laws the Communications Satellite Act. This established COMSAT, with Comsat's goal being to provide a global communications satellite system '*which will be responsive to public needs and national objectives, which* [was said to] *serve the communication needs of the United States and other coun*tries, *and which* [it was said] *will contribute to world peace and understanding*' (as per the Act. See discussions within Socol, 1977). Of course, today satellites are relevant to many operations used in a civilian and commercial setting, however satellites also serve as an invaluable aid to defence and therefore as a warfare tool too—which obviously begs the question in terms of contributing to a world of peace. Once again, this proactive national approach, shows the U.S. to be an earlier leader in space ventures which cross over into the realms of commercial and military (defence) activities.

From an international perspective the UN later established the Broadcasting Principles, governing the use by states of artificial earth satellites for international direct television broadcasting—this occurring in 1982.

2.3 The United States [Subsequent] Space Acts.... and Approaches

While the U.S. has amended its earlier Acts relevant to space, the nation has also enacted new legislation too. Particularly to note are the following Acts which address the move towards more commercial activities and in particular invite more involvement of the private sector.

2.3.1 The Commercial Space Launch Act

The Commercial Space Launch Act of 1984 is a federal law authorised to facilitate the private enterprise of the commercialisation of space and space technology. The legislation was enacted by the 98th Congressional session and signed by the 40th President of the United States, Ronald Reagan, on 30 October 1984, although over the years it has again been subject to amendments.

The Act was significant as it had the intention to acquire innovative equipment and services offered by entrepreneurial ventures from the information technology services, remote sensing technology and telecommunications industries. This clearly reinforcing the synergy and value of technology and communications within the space domain. Although written in terms of a commercial perspective there is little doubting of the value to not only civilian applications but defence also. While the Act also assigned the duties of overseeing and coordinating commercial launches, issuing of licenses and permits, and promotion of safety standards to the Secretary of Department of Transportation.

The Act, in essence, both recognised and reinforced the U.S. private sector as having the capability to develop commercial launch vehicles, orbital satellites and operate private launch sites and services. In fact, it invited the private sector to step forward, noting that '*private applications of space technology have achieved a significant level of commercial and economic activity, and offer the potential for growth in the future, particularly in the United States*' and did so by simplifying processes—such as the acquirement of licenses. That said the emphasis remained on the benefits to the U.S. and while foreign contributions were also welcomed the control lay firmly with the government in terms of specifying that a "United States citizen" means

(A) any individual who is a citizen of the United States;

(B) any corporation, partnership, joint venture, association, or other entity organized or existing under the laws of the United States or any State; and
(C) any corporation, partnership, joint venture, association, or other entity which is organized or exists under the laws of a foreign nation, if the controlling interest (as defined by the Secretary in regulations) in such entity is held by an individual or entity as described in the above two paragraphs.

This recognition, in terms of the contribution and involvement of the private sector, arguably paved the way for extending the concept of commercial space launches to include other activities that could be undertaken while in space.

2.3.2 U.S. Commercial Space Launch Competitiveness Act of 2015

The Commercial Space Launch Competitiveness Act was enacted in order '*to facilitate a pro-growth environment for the developing commercial space industry by encouraging private sector investment and creating more stable and predictable regulatory conditions,*' (as per the long title). While it also additionally tucked in at the end of the longer title, three key words, "*for other purposes*" also. This was a significant indicator of expansion from encouraging the private sector into the realms of commercial space launch activities into other areas—namely, 'Space Resource Exploration and Utilization.'

As identified previously (Fox, 2016, 2019) lobbying in the U.S. began shortly after the forming of Planetary Resources, Inc. The stated goal of the company was to expand Earth's natural resource base into space. On 19 March 2015 the Space Resource Exploration and Utilization Bill (SREU) was introduced for referral before the House of Representatives. The Bill was sponsored by Bill Posey a Republican Representative, and the aim was to approve the facilitation of commercial exploration and utilisation of space resources to meet 'the nation's needs.' (Fox, 2016). However, this was an unprecedented approach by any nation to take, not in terms of striving to meet national needs, given that '[t]he pursuit of resources drove the discovery of America and opened the West' (Eric E. Schmidt, Planetary Resources investor, cited in Fox, 2019) but in terms of seeking to take resources from an area beyond a national boundary—that is from (outer) space.

The Bill (as introduced) was subject to several amendments and much discussion, although it passed the respective stages relatively quickly (Fox, 2016, 2019) and was subsequently adopted through the implementation of the U.S. Commercial Space Launch Competitiveness Act—inserted at Title IV as the Space Resource Exploration and Utilization Act of 2015 (SREU Act).

In many ways, the SREU Act is rather short and brief in terms of content and it certainly does not reflect the level of discussion, scrutiny and controversy surrounding it. What it does however clearly state is that the U.S. government is extending rights to *it's citizens*—namely, '*to facilitate commercial exploration for and commercial recovery of space resources by United States citizens*' (at §51302 (1)). This part is far reaching and not specific as such but continues at (2) to identify that it enables this by discouraging government barriers to the '*development of….. commercial explorations for and commercial recovery of space resources;*' while, at point (3), reference is further made to *promoting* the rights of U.S. citizens '*to engage in commercial exploration and commercial recovery of space resources.*' At point (2) acknowledgement is made to this activity being consistent with international obligations of the U.S. and at point (3) while this is reiterated, reference is perhaps arguably contradictory though in terms of these activities being '*free from harmful interference.*' This is questionable in terms of the interpretation, and largely centres around the legitimacy of the U.S. to actually undertake resource 'recovery' or exploitation from space.

In this regard, space resources are not specified or defined at this point, although the paragraph earlier does refer to, albeit, a brief definitions section (§51301). Here the following is stated:

- That an asteroid resource: '*means a space resource found on or within a single asteroid.*'
- While a space resource more generally means, '*an abiotic* resource in situ in outer space.' This is interpreted in terms of a non-living entity but can be more extensive in terms of any ecosystem factor, attribute, component, or constituent that is devoid of life. Therefore, it can be interpreted as quite significant in terms of, although non-living, it is, however, part of an ecosystem that shapes its environment (National Geographic).

A definition is also provided in terms of clarifying who/what a citizen of the U.S. means, which largely mirrors the above, although reference is made specifically to Title 51. National and Commercial Space Programs § 50902 definition.

Perhaps the most specific and therefore controversial section is the again rather short paragraph referring specifically to 'Asteroid resource and space resource rights' (§51303). In its entirety it says,

> A United States citizen engaged in commercial recovery of an asteroid resource or a space resource under this chapter shall be entitled to any asteroid resource or space resource obtained, including to possess, own, transport, use, and sell the asteroid resource or space resource obtained in accordance with applicable law, including the international obligations of the United States.

The statement is both brave and overtly controversial. The U.S. certainly did not hide its intentions which it was clearly stating, by proactively staking a claim to the wealth potential above the Earth through commercial space resources, mining and property ownership. Arguably, this action was to indicate the next phase of globalisation and expansion out of Earth's orbit into asteroid'*sation,*' or 'aster*isation*' albeit, potentially of a monopolistic nature, which calling into question—humankind's rights vs. profit and commercialisation (Fox, 2016). Unlike the 1958 *National Aeronautics and Space Act,* the *Space Resource Exploration and Utilization Act* clearly came after the OST which the U.S. is a signatory too. While the OST fails to specifically mention mining or other commercial activities relating, more broadly, to resource extraction, it does refer to, within Article II, that '*Outer space, including the Moon and other celestial bodies, is not subject to national appropriation by claim of sovereignty, by means* *of use or occupation, or by any other means.*' It is, arguably, largely down to the potential to mine and extract resources from outer space that this sentence has, more recently, been subject to various interpretations (see related Chap. 5 discussions). The controversy and contention stemming from the U.S. SREU Act is that this is viewed by many as an extension of U.S. sovereignty and hence the creep of nations laws into (outer) space (Fox, 2022). It applies a very capitalist approach that appears at odds with the concept of a global common idea or common heritage of humankind approach (Fox, 2016, 2019, 2022).

Space is not the only domain though that the U.S. have resisted this stance being taken on, that is—in terms of restricting, what the U.S. sees as its right to commercially access and prosper through utilising (including obtaining resources and mining) common areas. In this way, there are clear parallels and lessons to be drawn from the deep sea and the seabed, as well as previous international space treaties—such as the Moon Agreement (MA) which the U.S. never signed or ratified (Fox, Chapter 5; Fox, 2016, 2022). However, the U.S., in its defence, in relation to mining of the seabed, have levied criticism, not against the principle of benefit sharing but the restrictions put in place and the requirement to share technology alongside the redistributed of wealth mechanism (Fox, 2022). This said, the United Nations Convention on the Law of the Sea (UNCLOS) is more specific than the OST or the Moon Treaty in terms of going some way to explaining the process for sharing acquired wealth. There is also an additional Implementation Agreement (1994) as well as an International Seabed Authority (ISA), created by UNCLOS, which, it is said, must arrange for the '*equitable sharing of financial and other economic benefits derived from* [deep-sea mining—DSM] *activities In the Area*' (UNCLOS—Articles 140 & 160). Although, even regarding DSM, it is said many questions still remain (Wilde et al., 2023), while the OST (and MA) lack even this depth of content and hence remain open to varying interpretation and thus, debate.

Academics remain divided (Fox, 2016). Ram Jakhu speaking shortly after the Bill was adopted said the overarching purpose of the OST leads to the conclusion that '*there really shouldn't be any private property rights in outer space.*' But should and won't are obviously subject to a morale perspective. His view being, '*that natural resources [in space] should not be allowed to be appropriated by anyone—states, private companies, or international organizations.*' While Ricky Lee (an Australian lawyer) argued, '*... the idea that commercial use of space resources is prohibited by the Outer Space Treaty... is quite simply absurd*' (CBC.ca). The U.S. continues to contend that this is permitted under the more open terms of the following cited sentence, 'Outer Space, including the Moon and other celestial bodies, shall be free for exploration and use by all States.' However, Fox writing shortly after the SREU Act and tracing the Bills steps also offers that this does not translate through to meaning the same as assuming possession of a quantity of resources mined on a large scale for commercial purposes. In other words, '*nowhere within the letter or OST is reference made to mining for commercial purposes and permission granted for this purpose*' (Fox, 2016). However, without this being specifically stated, similar to the U.S., it has also been assumed that it is a permitted action. While Blount and Robinson have argued that '*Outer Space Treaty was written in contemplation of* *innovation.*' Hence, as identified, perhaps the '*debate isn't about what it says. It's about what we want it to say*' (cited in Fox, 2019). Certainly the U.S. has continued to interpret that the OST does not prevent mining in outer space, largely because it represents their best interests to do so.

The SREU Act clearly provides that a U.S. citizen is granted rights to a space, or asteroid, resource that they mine. Hence, interpreted as the U.S. (as a nation) accepts that the property (resource) obtained would become owned by the person mining or *recovering* such a resource. In this way the U.S. arguably sought a national approach that addressed concerns expressed by 'stakeholders' over the 'legal and regulatory uncertainties … impeding their development' and that 'threaten[ed] to disrupt their continued investment and eventual activities in outer space' (Fox, 2016, discussions relating to the SREU Bill). Yet, this seems at odds with the disclaimer added to the subsequent Act in terms of expressing the U.S. opinion in relation to 'Extraterritorial Sovereignty, whereby it is said,

> It is the sense of Congress that by the enactment of this Act, the United States does not thereby assert sovereignty or sovereign or exclusive rights or jurisdiction over, or the ownership of, any celestial body.

This inclusion, perhaps, tried to address some of the concerns in terms of a potential '*space grab,*' or claim to a celestial body—yet, this, again, is really subject to interpretation. While saying that it would not make a sovereign claim of ownership to a jurisdiction, such as a celestial body or claim any exclusive rights to a space mass, it does, nevertheless, still exert ownership of property acquired or removed. reiterates the belief that it is acting in conformity with the OST (2016). Fox in 2016 questioned how this 'concept' would stand up to '*external scrutiny, and will be viewed by the international community, which* [she argued,] *is inherently more inclined towards treating such 'assetsas benefits for the whole of mankind' and not [for] a select few, where inevitably the national country would also take a share in the acquired wealth.*' Debatably, international intention of what is, should be, permitted in space, in relation to asteroids and other celestial bodies and resource extraction, is better detailed and defined in the UN Moon Agreement—yet, this Treaty has not been widely accepted. This therefore leaves the OST as the only binding international treaty on the topic; however, the question remains as to the applicability to space mining and resource extraction. This primarily focussing remains on whether 'exploration and use' includes space mining, and whether the provisions apply to private entities (Chap. 5).

2.3.3 The Artemis Accords

Contrary to the U.S. opinion, other nations, Russia and China in particular, have expressed the belief that the OST prohibits resource extraction and ownership (Byrd, 2022). This disagreement has ultimately sparked discussions in the United Nations Committee on the Peaceful Use of Outer Space, with Russia arguing that all space resource laws should be made by the Committee. Academics coming later into discussions though have sought to, perhaps, not argue so much on the legitimacy of space mining, rather on the concept of benefit sharing (Deplano, 2023). These thoughts and discussions no doubt emanating from the later actions of the U.S. in terms of the Artemis Accords, which as predicated would happen (Fox, 2022), has led to other academic voices approaching the table with further thoughts on a more pooled approach. The Artemis Accords relates to "Principles for Cooperation in the Civil Exploration and Use of the Moon, Mars, Comets, and Asteroids for Peaceful Purposes." Although, rather than being an international approach—they remain a State led, U.S. initiative. According to the U.S., the Artemis Accords aims to simply develop common grounds for space resources (Fox, 2022). Yet, these 'Accords' are also contentious and remain disputed by other nations—such as Russia and China. There are, in essence, are a '*non-binding set of principles designed to guide civil space exploration and use in the twenty-first century*' (U.S. Department of State).

They are co-led for the United States by the Department of State and NASA and were launched on 13 October 2020 with Australia, Canada, Italy, Japan, Luxembourg, the United Arab Emirates and the United Kingdom joining the United States. Other nations have since joined, including France and Germany which initially expressed some reluctancy (arguably concerns) before doing so (Fox, 2022). As of May 2025, there were 55 signatories. While the U.S. states that it aims to 'continue bringing together nations with a common vision for peaceful, sustainable, and transparent cooperation in space' (U.S. Department of State). And, according to the U.S. State Department, the Artemis Accords represent '*the best of multilateral leadership in civil space diplomacy and bring together a diverse set of nations with a shared vision of peaceful space cooperation. [Identifying that the] United States welcomes nations who share this vision to join us.*' To date neither China or Russia have signed up to these principles.

The U.S. continues on the other hand, to reaffirm that the Artemis Accords (hereafter the Accords) remain within the Outer Space Treaty's goals, identifying that the essence is based on cooperative activities, which should be exclusively for peaceful purposes and, therefore are in accordance with international law. Key sections of the Accords speak to not only peaceful purposes (section 3) but transparency (section 4), interoperability (section 5), emergency assistance (section 6), registration of space objects (section 7), release of scientific data (section 8), deconfliction of activities (section 11), orbital debris (section 12), alongside, sections on space resources and the protection of heritage (sections 10 & 9 respectively).

Hence, the Accords extend beyond resources and mining, with the principles establishing a common political understanding regarding mutually beneficial practices in the exploration and use of outer space. This noticeably being identified as including activities conducted in support of NASA's own Artemis Program. This 'Program' stretching also beyond resources and mining to going back to the Moon and exploring the Moon for '*scientific discovery, technology advancement*, [as well as learning] *how to live and work on another world as we prepare for human missions to Mars*.' While NASA reinforces the collaborative aspects that include commercialisation and working with international partners, including to '*establish the first long-term presence on the Moon*' (NASA).

In many ways though, the Artemis initiative (the Program and Accords) emphasise the political (astro-politcs) dimension of space, largely based on a national rather than international approach and one that aims to put and keep the U.S. firmly at the forefront. While it might be about collaboration, it is also built upon the realisation that no one nation is able to finance such an ambitious mission and that it requires commercial buy-in and international support—which is based upon both finance and also political agreement, namely including the U.S. taking the lead, supported by its 'friends' or, 'agreeing nations'. It is therefore based largely upon risk management, however defined, in terms of intersected financial risks and liabilities but particularly political risks and the realisation of the need to keep, not only world, but now space peace, too. This is echoed through the clear reference to the need for deconfliction of various activities—including mining and hence space resources. In. fact, the preamble identifies clearly that, the Accords,

> "RECOGNIZ[e] the global benefits of space exploration and commerce;" while,
>
> "ACKNOWLEDGING a collective interest in preserving outer space heritage."

Given the range of the sections (as above) and hence the remit, the Accords remain an ambitious set of loose 'Principles,' which arguably lack the depth needed to ensure peace is maintained in the execution of the Artemis Program. While the mechanism for deconflicting tensions is largely based on reaffirming earlier affirmations as per the OST and United Nations Guidelines for the Long-term Sustainability of Outer Space Activities adopted by the COPUOS in 2019. Again, given the discussions in this chapter (as well as Chap. 5) these remain subject to various interpretations, which speak little of achieving consensus to peaceful activities being achieved in space. It is noticeable that the section on deconfliction of activities comes after the section on resources, which also refer to the woolly guidance within the OST, namely that:

> The Signatories emphasize that the extraction and utilization of space resources, including any recovery from the surface or subsurface of the Moon, Mars, comets, or asteroids, should be executed in a manner that complies with the Outer Space Treaty and in support of safe and sustainable space activities. The Signatories affirm that the extraction of space resources does not inherently constitute national appropriation under Article II of the Outer Space Treaty, and that contracts and other legal instruments relating to space resources should be consistent with that Treaty.

The Accords offer nothing in terms of advocating how, and by what mechanism, any resources should be shared, which would no doubt remain subject to a commercial agreement, again drawn up or led by the U.S. in terms of a national document.

The term space resources in the Accords also lacks definition, but this said, it is likely interpretation would be inconformity of the earlier said U.S. Acts (Sect. 2.3.2 of this Chapter).

The mention of heritage is also an interesting inclusion within the Accords, given that very few nations have actually ventured into space, the heritage of activities, at least, remains predominately centred around the U.S. and Russia (formally the Soviet Union). While it 'could be' construed that this section is actually centred around humanities heritage to space—the U.S. has, however, confirmed that this 'heritage' is the relatively short period of time relating to outer space activities, which they consider *'to comprise historically significant human or robotic landing sites, artifacts, spacecraft, and other evidence of activity on celestial bodies.'*

While the U.S. method might remain both disputed and contentious, it, nonetheless, remains in line with the U.S. approach in terms of taking a definitive lead to advance space pursuits, albeit, arguably, while taking a largely U.S. centric approach. The fact that the nation has done so highlights the number of gaps that exist in international law.

3 Other National Approaches to the International '*Global Commons*' of Space!

The U.S. is no longer alone in introducing national space laws, with the UN actively encouraging nations to do so (Chap. 5). This said, the U.S. remains a minority nation in terms of States introducing very focussed resource centred space laws, with the UN being somewhat silent in terms of endorsing or disputing some of these approaches.

3.1 National Space Mining and Resource Approaches

Nations that have followed a similar approach to the U.S. in terms of domestic law giving private property rights over extracted space resources to its citizens include Luxembourg, the United Arab Emirates and Japan (Pons, 2021).

3.1.1 Luxembourg

Luxembourg introduced their space mining law in 2017 (Luxembourg Mining Law—LML) and then created a national space agency in 2018 (Luxembourg Space Agency—LSA). While the LML does not define space resources, reference is made within the explanatory document, published with the draft law, to the U.S. definition of space resources (Cheney, 2019). Hence the U.S. law can be verified as creating a lead or precedence in this regards.

It starts very decisively by stating that space resources are capable of being owned and goes on to explain the licensing process for space mining. Although it also identifies that the licensed activities must be done in accordance with Luxembourg's international law obligations. This, as always, remains the crux of debate in terms of the OST and it being confirmed or agreed as to the legitimacy of pursuing abstraction of resources from space.

In a similar vein to the U.S. reasoning for their law, it has been identified by academics that the main drive for Luxembourg was also to provide '*legal certainty as to the ownership of minerals and other valuable space resources identified in particular on asteroids*' (Feichtner, 2019). Yet, it is evident that Luxembourg lacks the direct flight experience and, perhaps indisputably, the same level of space expertise as the U.S. while it is clearly aspirational in terms of space activities, including resource acquirement and mining. However, one primary incentive may have been the heavy investment (circa $30 million) into the U.S. company Planetary Resources Inc. (Coldeway, 2016) and the potential opportunity to provide an alternative base for space mining operations to commence from, potentially with a lower cost license, or even easier licensing system being in place. In this way, arguably, nations will compete in being the dominant lead, or even launching state when the opportunity to extract, mine or obtain space resources become feasible.

In terms of the LML it is stated that, "*The authorisation for a mission shall only be granted if the applicant is a public company limited by shares (société anonyme) or a corporate partnership limited by shares (société en commandite par actions) or a private limited liability company (société à responsabilité limitée) of Luxembourg law or a European Company (société européenne) having its registered office in Luxembourg*" (Article 4).

This said, it is questionable whether the investment into Planetary Resources was a wise one to make, given that it later encountered difficulties securing further funding (Foust, 2018). The company, later that year, was acquired by ConsenSys (ConsenSys, 2018).

3.1.2 The United Arab Emirates (U.A.E.)

In December 2019, the U.A.E. passed Federal Law No. 12 on the Regulation of the Space Sector (U.A.E.—RSS). This law covers various regulated activities in space (as per Article 4). This includes:

- Space Resources exploration or extraction activities.
- Activities for the exploitation and use of space resources for scientific, commercial or other purposes.

Article 3 states the application of the Law as follows:

"*The provisions of this Law shall apply to Space Activities and other Space Sector-related activities that are carried out as follows:*

1. In the State's territory or the State's establishments outside the State's territory.
2. From ships or aircraft registered with the State or space objects registered by the State.
3. By persons who hold the nationality of the State, or companies that have a headquarters in the State."

In the definition section it defines space resources as '*[a]ny non-living resources present in outer space, including minerals and water*' (Article 1). Rather than referring to licenses the RSS refers to permits, whereby it is identified that the Council of Ministers is required to generally address how the permits will be regulated (Article 14), while permits are not allowed to be transferred except with prior approval (at point 7).

Though academics have argued that the law does not directly stipulate that space resources can be exploited—rather, it only includes in its list of regulated space activities (DePagter, date unknown), this said, this is contended as Article 18 clearly refers to the exploitation of space resources as well as their '*acquisition, purchase, sale, trade, transportation, [and] storage*'.

The RSS also includes liability provisions which refer back to international claims against the State and, it is, furthermore, identified that the authorisation 'of the Permit to others does not exonerate any Operator from any obligation or liability established prior to the date of the assignment' (Article 14 point 6).

3.1.3 Japan

Japan enacted legislation on the Promotion of Business Activities for the Exploration and Development of Space Resources (herein the Japanese Act or the Act) in 2021. The purpose of this Act is said to ensure the accurate and smooth implementation of conventions concerning the development and use of outer space and includes to promote business activities for the exploration and development of space resources by private business operators, by establishing special provisions for license under the provisions of the Space Activities Act. This includes specifying the rules for the acquisition of ownership of space resources and other necessary matters concerning such activities, with respect to the exploration and development of space resources, in accordance with the basic principles of the Space Act (Suzuki, 2021).

The Act provides a clear set of definitions, as follows: namely,

(i) "space resources" means water, minerals and other natural resources that exist in outer space, including the Moon and other celestial bodies.
(ii) "exploration and development of space resources" means any activities listed in any of the following subitems (excluding those conducted exclusively as scientific research or for the purpose of scientific research):
    (a) examination of the existence of space resources that contribute to the mining, extraction and other similar activities specified by Cabinet Office Order (referred to as "mining, etc." or "mined, etc." in (b) and Article 5); and
    (b) mining, etc. of space resources and related processing, storage and other acts specified by Cabinet Office Order (Article 2).

While Article 6 relates to ownership of resources, confirming that '*[a] person who conducts business activities related to the exploration and development of space resources shall acquire the ownership of space resources that have been mined,* etc. *in accordance with the business activity plan pertaining to the license,* etc. *for the exploration and development of space resources, by possessing said space resources with the intention to own.*'

Similar to the other domestic laws on the same, Japan reiterates that attention is to be paid not to hinder the sincere implementation of treaties and other international agreements Japan has concluded, which, although not specifying, would include the OST (Article 6). Article 7 also reaffirms that this Act seeks to apply an international harmonised system on the exploration and development of space resources '*in collaboration with foreign governments by cooperating with international organisations and other international frameworks.*' Yet, of course, currently this does not exist and, hence, has not been agreed. While it also speaks of sharing information and coordinating private business activities, which is perhaps also an aspiration from a global perspective.

The Act also details the licensing requirements for application by persons 'to conduct space activities are those who want to launch a spacecraft from Japan's land or those who want to control a spacecraft and have the spacecraft control facility within the boundaries of Japan' (DePagter). While, similar to the U.A.E.'s law, permits cannot be transferred without prior approval from the prime minister.

3.2 Other Commentary

The action by the U.S., in terms of its domestic legislation, may yet see other nations following suit in due course, while the ideal must surely be an international agreement being put in place. However, as world tensions increase (see related discussions in Chap. 5) the likelihood is that this is becoming less likely to achieve (Fox, 2019, 2022).

In 2021, the Legal Subcommittee of UNCOPUOS established a Working Group under the agenda item 'General exchange of views on potential legal models for activities in the exploration, exploitation and utilisation of space resources.' This initial 'Group' was later renamed, in 2022, as the Working Group on Legal Aspects of Space Resource Activities. It has developed a five-year working plan which is scheduled to conclude in 2027. This will see a report and a set of initial recommended principles for space resource activities for consideration by UNCOPUOS. The aim is for a possible adoption by the United Nations General Assembly as a dedicated resolution or other suitable action (UN—Space Resources Activities Group).

Although some scholars advocate that an alternative multilateral approach is perhaps advantageous and favourable (Rabitz, 2023), it potentially remains divisive, leading to divisions, not only on Earth but also in Space (Fox, 2022).

Meantime collaboration is also being undertaken by China and Russia with their project relating to establishing an International Lunar Research Station, which proposes a sustained human presence on the Moon that will also likely require in-situ resource extraction and usage (see discussions within Chap. 5 and Zwart et al., 2023).

4 Conclusion

There can be little doubting of the impact the U.S. has had in terms of the development of space activities and indeed it's foresight in terms of a legislative development that predates the United Nations and, hence, a coordinated international approach. This later being hastened by the U.S. competitiveness with the Soviet Union. Today, that competitiveness still continues, not only between the U.S. and Russia, but arguably globally, which now includes new private players and entities being given centre stage, albeit under the auspices of a nation or nations which seek to prosper both financially and also politically, in terms of dominance, not only in (outer) space but on Earth.

The U.S. was a leader in seeking to exploit outer space and clearly has its attention turned to what lies within it, particularly the riches of space resources. Again, it was forward-thinking in this respect, initially through its national, and very U.S. centric, domestic legislation—with the Space Resource Exploration and Utilization Act, coming within the U.S. Commercial Space Launch Competitiveness Act of 2015. These Acts overtly stating the U.S. intention in terms of commercial practice, wherein it was inviting and actively seeking the involvement of the private sector. Space invariably costs money and a hefty initial investment to exploit; and, driven by the thirst of private actors and the corporate sector (such as Planetary Resources), it was quick to enact legislation that would drive the possibility of resource acquirement and even mining in space. Although, as discussed, this approach was not without its critics, many questioning the compatibility of its actions in terms of the acceptance of national legal agreements with international treaties, particularly the OST, which the U.S. remains a signatory/ratifier of. However, due to the impreciseness of the language of the OST and the inability to specifically identify what activities are allowed, or not allowed in space, a lacuna exists, or even black hole, in terms of nations' interpretations. Although, conversely while most scholars agree that the Moon Agreement would prohibit space mining, this Treaty lacks signatories and ratification, which questionably calls into play how enforceable it would be in international law. Certainly the U.S. has not agreed to it, which invariably leaves the OST as the only binding international treaty on the topic and the Magna Carta for space (see Chap. 5) but with this gaping hole, nations (like the U.S.) are taking their own initiatives to plug it.

Yet, perhaps due to the costs associated with resource exploitation in space, and likely also in a further attempt to show the compatibility of its domestic national approach with the OST, the U.S. widened its direction to include the international community—this coming in the form of the Artemis Accords. The Accords though being a set of principles, and largely driven by the U.S.'s wider Artemis Program ambitions, were initially signed and agreed by only eight nations on the 13 October 2020. The nations, joining the U.S., at that time, being Australia, Canada, Italy, Japan, Luxembourg, the United Arab Emirates and the United Kingdom. Although the membership to this informal pact has grown, it is interesting to note that out of the original signatories, three (excluding the U.S.) have also developed national (domestic laws) which extends reach into the domain of outer space and, hence, space resources. The legislation in Luxembourg and the UAE law preceded the Artemis Accords, so perhaps it is not surprising to see them as initial supporters of the Artemis Accords, particularly given also that Luxembourg placed an investment into the U.S., via Planetary Resources. Although, arguably, the initial surge of actions by the U.S. in passing its domestic law has not been carried over to the ability to advance space mining and make this a viable reality. That said, given the wealth that lies above us, it has to be concluded that this will occur and when it does, there will likely be challenges. Certainly, national actions will be further questioned, and conflicts might yet occur, on not only a domestic but international stage—which could be extended from the Earth domain into outer space.

References

Artemis Accords: National Aeronautics and Space Administration (NASA), 'The Artemis Accords: Principles for Cooperation in the Civil Exploration and Use of The Moon, Mars, Comets, and Asteroids for Peaceful Purposes'. www.nasa.gov/specials/artemis-accords/index.html

Artemis Program (NASA): https://www.nasa.gov/humans-in-space/artemis/

Blount, P. J., & Robinson, C. J. (2015). One small step: The impact of the U.S. commercial space launch competitiveness act of 2015 on the exploration of resources in outer space. *NCJL & Tech., 18*, 160. Available at: https://scholarship.law.unc.edu/ncjolt/vol18/iss2/1

Broadcasting Principles Governing the Use by States of Artificial Earth Satellites for International Direct Television Broadcasting in 1982. General Assembly resolution 37/92 of 10 December 1982 (United Nations).

Byrd, L. C. (2022). Soft law in space: A legal framework for extraterrestrial mining. 71 Emroy L. J. 801.

CBC News. (2015). *U.S. space-mining law seen leading to possible treaty violations*. 26 November 2015. https://www.cbc.ca/news/science/space-mining-us-treaty-1.3339104

Cheney, T. (2019). There's no rush: Developing a legal framework for space resource activities. *J. Space L., 43*, 106–110.

Coldeway, D. (2016). Planetary resources mines Luxembourg for $28M in asteroid-hunting funds. https://techcrunch.com/2016/11/03/planetary-resources-mines-luxembourg-for-28m-in-asteroid-hunting-funds/

Commercial Space Launch Act (U.S.) Commercial Space Launch Act (U.S.) "H.R. 3942 — 98th Congress: Commercial Space Launch Act." www.GovTrack.us. 1983. July 17, 2025 https://www.govtrack.us/congress/bills/98/hr3942. And the, Space Resource Exploration and Utilization Act of 2015 (H. R. 1508).

Commercial Space Launch Competitiveness Act 114th Congress (2015–2016). H.R.2262 - 114th Congress. (U.S.) 129 STAT. 704 PUBLIC LAW 114–90—NOV. 25, 2015. https://www.congress.gov/bill/114th-congress/house-bill/2262/text

Communications Act of 1934, 47 U.S.C. § 151 (U.S.).

Communications Satellite Act, 47 U.S.C. § 701(a) (1970) (U.S.).

Consensys: https://cdn.consensys.io/uploads/PLANETARY-RESOURCES.pdf

Deplano, R. (2023). Inclusive space law: The concept of benefit sharing in the outer space treaty. *International and Comparative Law Quarterly*. https://doi.org/10.1017/S0020589323000234

DePagter Morgan M. (date unknown) "Who Dares, Wins:" How Property Rights in Space Could be Dictated by the Countries Willing to Make the First Move. ONLINE *Comment* . CJIL Online 1.2. University of Chicago, The Law School.

de Zwart, M., Henderson, S., & Neumann, M. (2023). Space resource activities and the evolution of international space law. *Acta Astronautica, 211*, 155–162.
ESA—The European Space Agency: https://space-economy.esa.int/article/33/what-is-the-space-economy
Feichtner, I. (2019). *Mining for humanity in the deep sea and outer space: The role of small states and international law in the extraterritorial expansion of extraction. Leiden Journal of International Law, 32*, 255.
Foust, J. (2018) *Planetary resources revising plans after funding setback*. Space News.
Fox, S. J. (2016). Space: The race for mineral rights. 'The sky is no longer the limit.' Lessons from Earth. *Resources Policy, 49*, 165–178.
Fox, S. J. (2019). Policing mining: In outer-space greed and domination vs. peace and equity a governance for humanity! *Resources Policy, 64*. https://doi.org/10.1016/j.resourpol.2019.101517
Fox, S. J. (2022). Exploiting – land, sea and space: Mineral Superpower' In the name of peace: A critical race to protect the depths and heights. *Resources Policy, 79*.
Fox, S. J. (2023). The UK's '*appetite*' for Space: *An Increased craving! Journal of Air Law and Commerce, 88*(4).
Fox, S. J. (2024). Space tourism – Ensuring a safe and equitable framework. Lessons learnt from the U.S. and Aviation. *Journal of Air Law and Commerce, 89*(3). https://doi.org/10.25172/jalc.89.3.4
Japan: *Space Resources Act Enacted*, Libr. Cong., https://perma.cc/WS9J-QZKS
Luxembourg Mining Law: Loi du 20 juillet 2017 sur l'exploration et l'utilisation des ressources de l'espace [Law of 20 July 2017 On The Exploration and Use of Space Resources], art. 1–2, Lux. Space Agency.
Luxembourg Space Agency: https://space-agency.public.lu/en/agency/lsa.html. See press release: https://gouvernement.lu/en/actualites/toutes_actualites/communiques/2019/09-septembre/12-luxembourg-space-agency.html
National Aeronautics and Space Act of 1958, Public Law #85-568, 72 Stat., 426.
National Geographic: defining 'abiotic'. https://education.nationalgeographic.org/resource/resource-library-abiotic-factor/
New Space Economy Congress: https://www.newspacecongress.cat/en/the-congress/
Pons, J. (2021). *US, Luxembourg, Emirates and Now Japan Take the Lead to Exploit Space Mining*. Atalayar (28, July 2021). https://perma.cc/SH4S-YGVZ
Rabitz, F. (2023). Space resources and the politics of international regime formation. *International Journal of the Commons, 17*(1), 243–255. https://doi.org/10.5334/ijc.1274
Socol, S. K. (1977). *Comsat's first decade: Difficulties in interpreting the Communications Satellite Act of 1962*. https://digitalcommons.law.uga.edu/cgi/viewcontent.cgi?article=2324&context=gjicl
Suzuki, T. (2021). *Japan: Legal issues in space business in Japan – volume 2*. Baker McKenzie. https://perma.cc/FAQ7-XMU9
The Act on Promotion of Business Activities Related to the Exploration and Development of Space Resources (Act No. 83 of 2021).
Treaty on Principles Governing the Activities of States in the Exploration and Use of Outer Space, Including the Moon and Other Celestial Bodies (signed 27 January 1967, entered into force 10 October 1967) 610 UNTS 205 (Outer Space Treaty).
United Arab Emirates: Federal Law No. 12 on the Regulation of the Space Sector (RSS).
United Nations Convention on the Law of the Sea - opened for signature Dec. 10, 1982, 1833 U.N.T.S. 397 (entered into force Nov. 14, 1994), 21 I.L.M. 1261 (1982); And Agreement to the Implementation of Part XI (1994).
United Nations: UN – Space Resources Activities Group: United Nations, General Assembly, committee on the peaceful use of outer space, UN Doc A/AC.105/1260 (19 April 2022).
U.S. State Department: https://www.state.gov/bureau-of-oceans-and-international-environmental-and-scientific-affairs/artemis-accords

Vedda, J. A. (2016). *National space council: History and potential.* Center for Space Policy and Strategy Policy Paper. https://aerospace.org/sites/default/files/2018-05/NationalSpaceCouncil.pdf

Wilde, D., Lily, H., Craik, N., & Chakraborty, A. (2023). Equitable sharing of deep-sea mining benefits: More questions than answers. *Marine Policy, 151.*

Sustainability in Space: Challenges, Governance, and Future Prospects

F. Frank Alparslan, Al-Amin Dabo, Saeed Nayeri, and Amin Hosseinian-Far

Abstract This chapter explores sustainability in space. The chapter provides a sound overview of pertinent sustainability challenges, and the governance required for future space exploration. It also examines the issues such as space debris, its environmental impact, and highlights the importance of effective space traffic management. Moreover, implications of commercialising space are succinctly discussed, and pertinent ethical considerations are examined. The role of international communities in working together, thinking creatively, and adopting innovation are emphasised in this work. The chapter concludes with recommendations for a sustainable future in space, through which we stress the need to balance technological advancement with environmental and ethical responsibilities.

Keywords Space transport management · Space debris · Sustainability · Space commercialisation · Space colonisation · Environmental impacts · Space governance

F. F. Alparslan · A.-A. Dabo · S. Nayeri
Faculty of Business & Law, University of Northampton, Northampton, UK
e-mail: frank.alparslan@northampton.ac.uk

A.-A. Dabo
e-mail: al-amin.dabo@northampton.ac.uk

S. Nayeri
e-mail: saeed.nayeri@northampton.ac.uk

A. Hosseinian-Far (✉)
Sustainability Research Group, Hertfordshire Business School, University of Hertfordshire, Hatfield, UK
e-mail: a.hosseinian-far@herts.ac.uk

H. Jahankhani et al. (eds.), *Space Mining*, Space Law and Policy,
https://doi.org/10.1007/978-981-95-5543-7_7

1 Introduction

Space exploration has long been heralded as a pinnacle of human achievement, symbolising technological prowess and scientific ambition. However, the environmental ramifications of these ventures are increasingly under scrutiny. The emissions from rocket launches, debris from jettisoned components, and resource demands of associated infrastructure pose significant ecological challenges. Studies indicate that the impact of space exploration extends across terrestrial, atmospheric, and marine ecosystems, necessitating a holistic understanding of these effects to guide sustainable practices (Koroleva et al., 2018; Brown et al., 2024).

In light of these developments, this chapter provides an overview of areas that exacerbate sustainability in space. These include the surging challenge of space debris, and potential solutions, environmental contaminations associated with space launches, and the challenges and gaps associated with space transport management. The chapter also briefly discussed the notion of space commercialisation and colonisation. Furthermore, the chapter discussed the foundational international treaties, such as the Outer Space Treaty (UNOOSA, 1967), and the legal challenges associated with the expanding role of private entities in space exploration and utilisation. It seeks to analyse emerging issues in governance, including the need for sustainable development, equitable benefit sharing, and the role of soft law mechanisms in addressing contemporary challenges.

The chapter is organised as follows: Sect. 2 provides a succinct overview of current space debris landscape, challenges and risks associated with space debris, and a number of technologies that can potentially be found effective in mitigating the debris problem. Section 3 focuses on the environmental impacts of space launches and reviews a number of potential solutions. Section 4 examines space commercialisation and colonisation, followed by a review of space transport management in Sect. 5. Section 6 provides an overview of governance and legal frameworks associated with space and highlights a few recommendations. The Chapter is concluded in Sect. 7, summarising the key points, and future research pathways.

2 Space Debris

Space debris has emerged as a critical issue for the sustainability of space operations. It encompasses non-functional satellites, spent rocket stages, and fragments from collisions or explosions in orbit (Svotina & Cherkasova, 2023). The growing density of debris in Low Earth Orbit (LEO), Medium Earth Orbit (MEO), and Geostationary Earth Orbit (GEO) threatens operational satellites, human missions, and the future of space exploration (Shan et al., 2020). Since the launch of Sputnik 1 in 1957, the space environment has become increasingly congested. The continuous expansion of satellite constellations, driven by advancements in technology and reduced launch costs, has exacerbated the challenge of managing orbital debris (Nozawa et al., 2023).

Current estimates suggest that over 8800 metric tonnes of debris are circulating in Earth's orbits, with much of it concentrated in critical zones such as LEO (Maclay & McKnight, 2021). This accumulation has heightened concerns over the safety of space operations and the potential for catastrophic collisions. The issue is further complicated by the long lifespan of debris in orbit. Objects in LEO can remain for decades before re-entering Earth's atmosphere, while those in higher orbits, such as GEO, may persist for centuries (Hakima & Emami, 2018). This longevity increases the likelihood of collisions, which can generate additional debris in a chain reaction described by the Kessler Syndrome (Svotina & Cherkasova, 2023). Addressing this challenge requires innovative solutions and global collaboration to ensure the sustainable use of space for future generations.

Space debris threatens operational satellites and poses significant risks to human life during space missions. The International Space Station (ISS) has performed numerous collision avoidance manoeuvres to mitigate such dangers, which indicates the severity of the problem (Shan et al., 2020). With the growing reliance on satellites for communications, navigation, and scientific research, the stakes for effective debris management are higher than ever.

2.1 The Kessler Syndrome: A Cascade Effect

The Kessler Syndrome, introduced in 1978, describes a scenario where the density of debris in orbit reaches a critical point, subsequent to which a cascade of collisions renders certain orbital regions unusable for generations (Kessler & Cour-Palais, 1978; Svotina & Cherkasova, 2023). Historical events, such as the 2009 Cosmos-Iridium collision, provide real-world examples of this risk, which can generate thousands of new debris fragments and escalate the likelihood of further collisions (Shan et al., 2020). Projections indicate that without active interventions, the debris population will grow exponentially and can severely limit access to key orbital regions (Hakima & Emami, 2018). The concept of cascading collisions is exacerbated by the increasing number of satellites and the growing popularity of mega-constellations. Mega-constellations, consisting of thousands of satellites, significantly increase the risk of accidental collisions. As described by Nozawa et al. (2023), such constellations can create a feedback loop where collisions generate more fragments, which in turn increase the probability of subsequent collisions. This self-propagating cycle is a characteristic of the Kessler Syndrome and poses severe threats to both existing infrastructure and planned missions. Furthermore, the Kessler Syndrome's impact extends beyond operational disruptions. Debris collisions can generate high-velocity fragments that damage functioning satellites, leading to costly repairs or replacements. According to Forshaw et al. (2016), this risk necessitates a proactive approach to debris mitigation, including coordinated international efforts to remove large debris and prevent further fragmentation.

Considering the gravity of the situation, international regulatory bodies such as the United Nations Office for Outer Space Affairs (UNOOSA) emphasise the importance of sustainable practices. Collaborative measures, such as setting stricter end-of-life disposal guidelines and investing in Active Debris Removal (ADR) technologies, are crucial to mitigating the cascading effects of debris proliferation (Hakima & Emami, 2018).

2.2 Current Space Debris Landscape

As of 2020, the United States Space Surveillance Network (US-SSN) has catalogued over 23,000 debris objects larger than 10 cm, while millions of smaller fragments remain untracked but still pose significant risks (Shan et al., 2020). This number has risen to over 36,500 tracked pieces of debris larger than 10 cm as of 2022 (McClintock, 2023). LEO is particularly congested, with active satellites making up only a fraction of the objects in orbit. Approximately 95% of the tracked debris originates from fragmentation events or defunct spacecraft (Hakima & Emami, 2018).

The accumulation of debris has also led to an increase in collision avoidance manoeuvres by operational satellites. For instance, the European Space Agency (ESA) conducted 28 collision avoidance manoeuvres in 2020, a significant rise compared to previous years (Maclay & McKnight, 2021). These manoeuvres increase operational costs, and also highlight the growing difficulty of maintaining safe orbital pathways.

The majority of debris in LEO originates from historical fragmentation events. The destruction of the Fengyun-1C satellite in 2007 and the Cosmos-Iridium collision in 2009 collectively generated over 5000 trackable fragments, many of which remain in orbit today (Hakima & Emami, 2018). These events illustrate the long-lasting impact of individual incidents on the space environment. In addition to large debris, smaller, non-trackable fragments pose significant risks. These particles, while often under 10 cm in size, travel at high velocities and are capable of causing severe damage to operational satellites and space infrastructure (Zhao et al., 2020). The challenge of tracking and mitigating these smaller fragments remains an area of active research and technological development. Nevertheless, a number of mitigation technologies are currently being developed, which are outlined in the subsequent subsections.

2.3 Active Debris Removal (ADR) Technologies

2.3.1 Net-Based Capture

Net-based methods have attracted attention as effective tools for capturing debris. These systems involve deploying a flexible net tethered to a chaser satellite, capable of enclosing uncooperative targets (Zhao et al., 2020). A study by Shan et al. (2020) has demonstrated the viability of net-based systems, highlighting their ability to address large debris objects in a cost efficient manner. Moreover, advancements in net flexibility modelling have enhanced their performance, which can allow for more precise and effective debris capture (Shan et al., 2020).

2.3.2 Harpoon Technology

The harpoon system, as demonstrated in the RemoveDEBRIS project, uses high-speed projectiles to capture debris securely. This system has been successfully tested in simulated conditions and represents a promising method for larger, solid debris objects (Forshaw et al., 2016). The RemoveDEBRIS mission, launched in 2018 by Surrey Space Centre, successfully demonstrated techniques such as a net capture system and a harpoon to capture simulated debris (University of Surrey, 2019). While progress has been made in areas such as de-orbit sails and active debris removal demonstrations, scaling these solutions to address the vast number of existing debris remains a significant hurdle. Recent improvements in material strength and guidance systems have further increased the feasibility of harpoon-based debris removal, which make it an essential tool in addressing high-mass debris in orbit (Mark & Kamath, 2019).

2.3.3 Laser Ablation Techniques

Laser-based systems offer a non-contact solution for debris removal. Using directed energy for vaporising surface material, these systems create a reaction force that reduces the orbital velocity of the debris, which can then enable a controlled re-entry (Mark & Kamath, 2019). While still in developmental stages, laser ablation is considered highly scalable for small debris and has the potential to mitigate fragmentation risks during the removal process (Hakima & Emami, 2018).

2.3.4 Tethered Systems

Tethered systems, such as the Tethered Space Net Robot (TSNR), utilise a combination of flexible nets and controlled manoeuvring units to capture debris. This technology is particularly advantageous for uncooperative targets and long-distance captures (Zhao et al., 2020). Innovations in tether control dynamics have enhanced the deployment accuracy and post-capture stability, that ensures a reliable operation under varied orbital conditions (Zhao et al., 2020).

2.3.5 Robotic Arm Technology

Robotic arms have been developed as precise tools for capturing and manipulating debris. These systems are equipped with multiple degrees of freedom; This allows them to grasp objects of varying shapes and sizes (Hakima & Emami, 2018). Robotic arms are particularly effective for de-orbiting large defunct satellites and for servicing operational spacecraft, providing a dual-purpose utility in active debris management.

2.4 Economic Implications

As referred to earlier, the economic impact of space debris is noteworthy; Svotina and Cherkasova (2023) highlight that a region like LEO, which are critical for communications and Earth observation, could become prohibitively expensive to utilise due to the increased insurance and operational costs associated with high debris densities. As debris continues to accumulate, the cascading effect predicted by Kessler and Cour-Palais (1978) could permanently compromise these vital orbital zones.

Uncontrolled debris growth is projected to reduce global GDP by approximately 1.95% in the long term if no mitigation measures are implemented (Nozawa et al., 2023). The loss of critical satellite services, such as communications and navigation, would disrupt multiple industries (Hakima & Emami, 2018). The financial burden of managing debris-induced damages and collision avoidance manoeuvres has already escalated for satellite operators. For instance, studies estimate that satellite collision avoidance strategies incur significant operational costs, often diverting resources from innovation and new missions (Maclay & McKnight, 2021). Furthermore, sectors such as finance, transportation, and agriculture, which rely heavily on uninterrupted satellite services, face multiple risks associated with debris proliferation. Disruptions in Global Positioning System (GPS) signals alone could lead to widespread economic losses, with cascading effects on global supply chains (Van Camp & Peeters, 2022).

As the risk of collisions rises, insurers demand higher premiums to cover potential losses, further driving up the cost of satellite operations, and subsequently the insurance premiums (Hakima & Emami, 2018). This financial strain disproportionately affects smaller operators and emerging economies, limiting their participation in the global space economy (Nozawa et al., 2023). Mitigation measures, such as investing in ADR technologies, can potentially address environmental concerns, and also offer long-term economic benefits. Implementing systems such as laser ablation or tethered nets can reduce future collision risks and stabilise insurance costs (Forshaw et al., 2016; Zhao et al., 2020). Additionally, international collaborations on debris removal initiatives could distribute costs and enhance technological development, fostering a cooperative approach to space sustainability (Mark & Kamath, 2019).

3 Environmental Impacts of Space Launches

The environmental consequences of space missions are multi-layered, ranging from chemical emissions during rocket launches to the deposition of hazardous materials in terrestrial and marine ecosystems. This section explores the environmental footprint of space activities, including the challenges related to traditional propulsion systems and the potential of innovative solutions such as zero-emission propulsion systems and on-orbit servicing technologies.

3.1 Chemical Emissions and Atmospheric Effects

Rocket launches release significant quantities of greenhouse gases (GHG) and particulate matter directly into the atmosphere. These emissions occur across multiple layers of the atmosphere, with implications for the ozone layer and global climate. Solid rocket propellants, such as those used in Space Shuttle launches, release chlorine, which depletes stratospheric ozone. Comparatively, liquid propellants such as kerosene and liquid oxygen produce fewer ozone-depleting substances yet still contribute to atmospheric pollution (Dallas et al., 2020). Research indicates that emissions from launches can introduce nitrogen oxides and aluminium particulates into the stratosphere, which can then accelerate chemical reactions that degrade ozone molecules (Brown et al., 2024). Moreover, the potential for mesospheric cloud formation from rocket exhaust highlights the need for further studies on the cumulative impacts of increasing launch frequencies.

Existing literature have highlighted the extent of black carbon emissions produced by rocket engines, which significantly influence radiative forcing in the stratosphere. Black carbon particulates absorb sunlight, warming the upper atmosphere and potentially altering weather patterns on a global scale (Lonsdale & Phillips, 2021). Additionally, the combustion of solid rocket motors generates alumina particles, which act as condensation nuclei, potentially affecting cloud microphysics and precipitation patterns (Koroleva et al., 2018). Liquid rocket propellants, although less harmful to the ozone layer than solid fuels, still present environmental challenges. For instance, kerosene-based propellants emit carbon dioxide, water vapour, and soot, all of which contribute to atmospheric warming and cloud formation (Sharapova et al., 2020). These emissions are particularly concerning, considering the projected increase in rocket launches to support satellite mega-constellations and space tourism (Brown et al., 2024).

The deposition of reactive chlorine species from solid rocket motor exhaust has been observed to create localised 'ozone holes' near launch sites. These phenomena, although temporary, indicate the immediate impact of rocket emissions on atmospheric chemistry (Dallas et al., 2020). Furthermore, nitric oxide released during launches can catalyse ozone destruction cycles, i.e., exacerbating the ozone depletion caused by other pollutants (Lonsdale & Phillips, 2021).

3.2 Ground-Level Environmental Contamination

Rocket stages that are jettisoned during launches often fall back to Earth, which then leads to residual propellants and debris. Studies at sites such as the Baikonur Cosmodrome have highlighted soil and water contamination by unsymmetrical dimethylhydrazine (UDMH) and nitrogen tetroxide, two commonly used rocket fuels (Koroleva et al., 2018; Sharapova et al., 2020). Such contamination leads to long-term ecological damage, including altered pH levels and toxic accumulation in terrestrial ecosystems, which can have ripple effect on soil quality, agriculture, and so on. The impact on snow-covered regions is particularly severe, with pollutants such as nitrates and ammonium persisting in the environment. In areas with thinner snow cover or higher temperatures, the diffusion of contaminants is more noticeable (Sharapova et al., 2020). Moreover, the deposition of particulate matter from rocket exhaust contributes to the degradation of local flora and fauna. Vegetation in the vicinity of rocket landing zones often shows signs of stress and reduced growth, attributed to the toxicity of the chemicals involved (Dallas et al., 2020).

Ecosystems impacted by debris from liquid-fuelled rockets exhibit both chemical and mechanical disruption. Studies have reported increased concentrations of hydrocarbons and nitrates in soil samples collected from landing zones, which can persist for decades (Koroleva et al., 2018). Such pollutants create significant risks to groundwater quality and agricultural productivity in affected areas. For instance, hydrazine-based fuels can alter microbial activity in soil, which can then lead to imbalances in nutrient cycling and reduced fertility (Sharapova et al., 2020). Another concern involves the health implications for nearby human populations. Exposure to residual propellants such as UDMH has been linked to respiratory diseases and other health risks, which necessitates stringent safety protocols and mitigation strategies. These risks highlight the importance and the urgent need of developing eco-friendly alternatives to traditional propellants and designing recovery systems to minimise contamination (Brown et al., 2024).

3.3 Marine Impacts of Space Debris

Similar to land, the marine environment is increasingly affected by space debris from jettisoned rocket stages and satellite components. Debris falling into oceans disrupts ecosystems and contributes to pollution. The UK's marine impact assessment for small satellite launches highlights the risks associated with unchecked debris deposition, which includes potential toxic leaching from propellant residues (Lonsdale & Phillips, 2021). Moreover, propellant residues such as UDMH can lead to chemical contamination in seawater. Studies have shown that such pollutants, depending on their concentration and persistence, may bioaccumulate, causing disruptions in trophic levels and endangering biodiversity (Sharapova et al., 2020). Moreover, physical debris from launch stages often sinks to the ocean floor, where it creates hazards for benthic organisms and alters sediment composition (Dallas et al., 2020).

The geographic distribution of marine impacts is also noteworthy, as debris typically falls in international waters or exclusive economic zones (EEZs). This raises questions of regulatory accountability and necessitates international collaboration for debris management (Lonsdale & Phillips, 2021). Current legislative frameworks, such as the United Nations Convention on the Law of the Sea (UNCLOS), provide limited guidance on managing space debris in marine environments, which indicates a gap in the pertinent global governance (Brown et al., 2024). Efforts to mitigate marine impacts include the development of controlled re-entry technologies, that aim to reduce the likelihood of debris falling into sensitive marine areas. Additionally, initiatives such as ADR have been proposed to retrieve hazardous materials from oceans, though these technologies remain in their infancy and require substantial investment (Koroleva et al., 2018).

3.4 Innovative Solutions for Mitigating Environmental Costs

3.4.1 On-Orbit Servicing

Technologies enabling on-orbit repairs and refuelling of satellites reduce the need for repeated launches, which can ultimately reduce the environmental impacts. This approach aligns with broader sustainability goals in space exploration, as it maximises the utility of existing infrastructure and minimises waste. On-orbit servicing can significantly extend the operational lifetimes of satellites, hence reducing the demand for new launches. Additionally, advances in robotic technologies have enabled precision repairs in orbit, which can potentially reduce the risk of mission failure.

Recent developments have demonstrated the feasibility of in-orbit assembly and manufacturing, which could eliminate the need for some launches entirely. As reported by Dallas et al. (2020), such technologies have the potential to revolutionise space exploration by creating modular systems that can be upgraded or repaired in space. Modular systems with a similar concept to those used in Earth-based constructions (Gharehbaghi, 2017) can offer sustainable solutions for space systems. This modular approach can reduce the ecological footprint associated with launching entirely new systems. However, the implementation of on-orbit servicing faces technical and regulatory challenges. For instance, ensuring interoperability between servicing vehicles and client satellites requires standardised interfaces and communication protocols (Brown et al., 2024). Furthermore, regulatory frameworks shall address liability and ownership issues in the event of servicing failures or collisions, as highlighted by Lonsdale and Phillips (2021). Despite these challenges, the potential environmental benefits of on-orbit servicing make it a valuable consideration for sustainable space operations.

3.4.2 Zero-Emission Propulsion Systems

Zero-emission propulsion technologies, such as solar sails and electric thrusters, offer promising alternatives to traditional chemical rockets. Although their current use is limited to specific missions, their scalability holds potential for reducing the carbon footprint of space activities (Brown et al., 2024). One of the most promising zero-emission propulsion technologies involves the use of solar sails, which harness photon pressure for thrust. These systems produce no direct emissions, which makes them particularly attractive for long-duration deep space missions (Dallas et al., 2020). Similarly, advancements in ion propulsion systems—ion thrusters—provide a highly efficient and low-emission alternative to chemical propulsion (Koroleva et al., 2018).

The integration of zero-emission systems however entails a number of challenges. For instance, solar sail deployment requires precise control mechanisms to optimise trajectory adjustments, while ion thrusters demand significant power inputs (Sharapova et al., 2020). Nevertheless, the environmental benefits of these technologies justify continued investment and research. Emerging concepts such as electrodynamic tethers also show potential for reducing the environmental impact of space missions. Through interaction with Earth's magnetic field, these tethers can generate propulsion without fuel consumption, which has the potential to offer a sustainable solution for de-orbiting defunct satellites (Lonsdale & Phillips, 2021). However, practical implementation of such technologies requires overcoming technical barriers, including material durability and tether deployment mechanisms.

4 Space Commercialisation

Space commercialisation can be deemed as a significant transformation in humanity's approach to space exploration and utilisation. Moving beyond the government-led missions of the past, the private sector now plays an important role in driving innovation and expanding access to outer space (Hampson, 2017). This shift has opened the door to a wide range of economic activities, from satellite services to resource extraction and tourism (Mitchell, 2018). In line with this commercialisation effort, the establishment of bases on the Moon and Mars represents humanity's ambition to extend its presence beyond Earth. These bases, envisioned as research hubs and stepping stones for deep space exploration, can redefine the scope of human activity in the cosmos (Zuniga et al., 2015). Together, space commercialisation and extraterrestrial colonisation represent the twin pillars of a new era in space development. This section explores the interconnection between these domains, examining the economic, technological, and societal implications of this unprecedented expansion. Understanding the drivers, challenges, and future trajectories of these initiatives provides comprehensive insights for their transformative potential (Zubrin, 2019).

4.1 The Evolution of Space Commercialisation

The commercialisation of space began modestly with the deployment of communications satellites in the mid-twentieth century. Early examples, such as Telstar and Intelsat, were primarily funded and operated by governments or international consortia. However, the advent of privatisation and deregulation in the late twentieth century paved the way for the emergence of commercial space companies (Sturdevant, 2024). This transition was further accelerated by the declining cost of space launch technology, driven by innovations such as reusable rockets (Bushnell & Moses, 2018). Today, the space industry is a multi-billion-dollar sector encompassing diverse activities. Satellite communications and Earth observation services dominate the market, and provide critical capabilities for telecommunications, weather forecasting, and disaster management (Mitchell, 2018). At the same time, new markets are emerging, including space tourism, in-orbit manufacturing, and asteroid mining. The entry of private companies such as SpaceX, Blue Origin, and Virgin Galactic has transformed the industry, which have led to lower costs and greater accessibility to space than ever before (Maraš & Dangubić, 2022). The economic benefits of space commercialisation extend far beyond the industry itself. Space-based services underpin many aspects of modern life, from GPS navigation to global internet connectivity. Additionally, the commercialisation of space creates new job opportunities and stimulates technological innovation, with applications in fields as diverse as robotics, materials science, and artificial intelligence (AI) (Mitchell, 2018).

The success of space commercialisation and colonisation relies on a suite of advanced technologies that enable sustainable operations in challenging environments. Among these, reusable rockets have had the most transformative impact, dramatically reducing the cost of access to space. SpaceX's Falcon 9 and Starship systems exemplify the potential of this technology, enabling frequent launches at a fraction of the cost of traditional expendable rockets (Lubert, 2018). Habitat design is also a key area of innovation, with researchers exploring modular and inflatable structures that can be deployed and assembled on-site. These habitats must provide protection against radiation, temperature extremes, and micrometeorites while supporting the physical and psychological well-being of their occupants. Advances in robotics and AI further enhance the feasibility of extraterrestrial bases by enabling autonomous construction, maintenance, and resource extraction (Naser & Chehab, 2018; Wu et al., 2024).

4.2 The Development of Lunar and Martian Bases

Although the commercialisation of near-Earth space has attracted significant attention, the long-term vision of space exploration extends far beyond Earth's orbit. Establishing bases on the Moon and Mars is a cornerstone of this vision, which can both serve as platforms for scientific research and as stepping stones for humanity's eventual expansion into the solar system (Zuniga et al., 2015).

4.2.1 Lunar Bases: Gateway to the Solar System

The Moon is often considered the logical first step in extraterrestrial colonisation due to its proximity to Earth and relatively well-understood environment. Lunar bases would facilitate a wide range of activities, including resource extraction, scientific research, and the testing of technologies needed for deeper space missions. The presence of water ice in the Moon's polar regions is particularly significant, as it can be converted into oxygen, hydrogen, and drinking water, reducing reliance on Earth-based supplies (Spudis, 2016). NASA's Artemis programme is one of the most ambitious initiatives aimed at establishing a permanent human presence on the Moon. The programme envisions a series of crewed missions culminating in the construction of a sustainable lunar base by the late 2020s. Key to this effort is the Lunar Gateway, a space station orbiting the Moon that will serve as a hub for logistics, research, and crew transfer (NASA, 2021).

4.2.2 Martian Bases: Humanity's Next Frontier

Mars presents a more challenging yet equally compelling target for colonisation. The planet's similarities to Earth, including its day length and potential for resource utilisation, make it a suitable candidate for human settlement. However, the challenges of reaching and inhabiting Mars are numerous. Long travel times, harsh surface conditions, and high levels of radiation cause significant risks to both humans and equipment (Zubrin, 2019). Despite these challenges, visionary plans for Martian colonisation are already underway. SpaceX has outlined an ambitious roadmap for establishing a self-sustaining city on Mars using its Starship spacecraft. This vision includes the transport of large numbers of settlers and the development of in-situ resource utilisation (ISRU) technologies to support life on the Martian surface (Musk, 2017). ISRU systems are designed to extract and process local resources, such as water, regolith, and atmospheric gases, to produce life support consumables, construction materials, and rocket fuel. In parallel, NASA and other space agencies are conducting robotic missions to gather data and test technologies that will pave the way for human exploration (Görner et al., 2024).

4.3 Interconnections Between Commercialisation and Colonisation

Space commercialisation and the development of extraterrestrial bases are interconnected, with each domain providing critical support for the other. Commercial space activities generate the financial resources and technological innovations needed to support colonisation efforts. In turn, the establishment of lunar and Martian bases creates new markets for commercial space companies, which can then drive further investment and growth (Gottlieb, 2019). For instance, resource extraction on the Moon or asteroids could provide the raw materials needed for space manufacturing and can reduce the cost and complexity of constructing habitats and infrastructure on Mars. Similarly, commercial satellite networks could enable high-speed communication between Earth and extraterrestrial bases, which can potentially facilitate real-time coordination of operations. This symbiotic relationship highlights the importance of integrating commercialisation and colonisation efforts into a cohesive strategy for space development (Zubrin, 2019).

As humanity expands its footprint into space, ethical and legal questions become increasingly pressing. Who has the right to exploit extraterrestrial resources? How should the benefits of space exploration be distributed? What responsibilities do we have to preserve the pristine environments of other celestial bodies? These questions are at the heart of debates about the future of space commercialisation and colonisation (Forganni, 2023). The Outer Space Treaty of 1967 does not explicitly address issues such as resource ownership or private enterprise, which has left significant gaps in the regulatory landscape. Developing a comprehensive legal framework that balances commercial interests with ethical considerations will be essential to ensuring the responsible use of space (Chrysaki, 2020).

In the future, emerging markets such as asteroid mining and space-based energy production could unlock unprecedented economic potential (Simpson, 2014; Steffen, 2022). The future of extraterrestrial bases is equally promising, with plans for permanent habitats on the Moon and Mars becoming increasingly detailed and actionable, however, realising this vision will require sustained investment, international cooperation, and technological innovation (Naser & Chen, 2021). Ultimately, the convergence of commercialisation and colonisation reflects humanity's broader aspirations for the twenty-first century and beyond. By harnessing the resources and opportunities of outer space, we have the potential to address some of the most pressing challenges on Earth, from energy scarcity to climate change. At the same time, the pursuit of space exploration embodies the enduring human spirit of curiosity and adventure, inspiring future generations to reach for the stars.

5 Space Traffic Management

The surge in human activities in space, driven by technological advancements and decreasing launch costs, has opened new opportunities in communications, navigation, and Earth observation. However, as highlighted earlier, this growth has also presented significant challenges, including orbital congestion and the increasing risk of collisions (Muelhaupt et al., 2019). Space Traffic Management (STM) is another systematic effort to address these challenges by integrating technical, regulatory, and governance solutions. STM focuses on the safety, sustainability, and efficiency of orbital operations (Hilton et al., 2019; McClintock, 2023). With over 6900 operational satellites currently in orbit and thousands more planned, the ability to organise and govern space activities is now critical to the sustainability of space as a resource (McClintock, 2023). Unlike terrestrial domains, STM operates in an environment where vast distances, diverse trajectories, and the absence of physical boundaries create unique challenges. A single collision could trigger the Kessler Syndrome, making entire orbital regions unusable for decades or even centuries (Hilton et al., 2019), as also highlighted in earlier sections.

5.1 The Evolution of Space Traffic Management

Historically, space activities were dominated by state actors, primarily the United States and the Soviet Union during the Cold War. With relatively few spacecrafts and launches, collisions were rare, and formal traffic management was unnecessary (Ailor, 2020). Space operations during this period were governed by broad international treaties such as the 1967's Outer Space Treaty, which emphasised peaceful exploration and the shared responsibility of space activities. However, the twenty-first century has seen a dramatic shift. The commercialisation of space, by companies such as SpaceX, Amazon, and OneWeb, has revolutionised access to LEO. These companies are deploying thousands of satellites to provide global broadband coverage and other critical services, making LEO one of the most congested regions in space (Sorge et al., 2021). Mega-constellations, consisting of thousands of satellites, are crowding LEO and increasing the likelihood of collisions. Simultaneously, the proliferation of space debris from defunct satellites and past collisions has compounded the challenges of STM (McClintock, 2023).

As such, STM has evolved into a strategic necessity, entailing satellite tracking, collision avoidance, orbital slot allocation, and compliance with international regulations. Its scope is no longer limited to ensuring operational safety and now includes the preservation of orbital resources for future generations. Arnas and Linares (2023) highlight the importance of equitable access to orbital slots, particularly for emerging spacefaring nations, as a critical element of STM.

5.2 *The Components of Space Traffic Management*

Space Traffic Management (STM) includes several interrelated components that collectively enable the smooth functioning of space operations. Each component is vital to addressing the challenges created by the increasing congestion of Earth's orbital environment and ensuring the long-term sustainability of space activities:

- STM is based on tracking and monitoring space objects. This involves identifying and cataloguing all active and inactive satellites, debris, and other objects in orbit. Ground-based radar systems and optical telescopes have traditionally been the primary tools for tracking. However, the high volume of objects in orbit has necessitated the development of more advanced technologies (McClintock, 2023),
- Space-based sensors and AI algorithms now play a crucial role in providing real-time data on the position, velocity, and trajectory of space objects. AI systems can process vast datasets to improve the accuracy and speed of decision-making and reduce the risk of collisions (Pinto et al., 2020). Furthermore, international collaborations, such as the European Space Agency's (ESA) Space Surveillance and Tracking (SST) programme, are advancing global capabilities in space object monitoring,
- Predicting and preventing collisions is a critical component of STM. This involves analysing the trajectories of multiple objects to identify conjunctions, or close approaches between satellites or debris. Once a conjunction is detected, operators must calculate and execute manoeuvres to adjust the trajectory of their spacecraft, to avoid potential impacts (Sorge et al., 2021). Nevertheless, predicting the long-term trajectories of space objects is fraught with uncertainty due to factors such as orbital decay and gravitational perturbations (Khan & Ntantis, 2024).
- Although collision avoidance is routine for larger satellites with manoeuvring capabilities, it becomes increasingly challenging for smaller satellites, particularly those in dense orbital regions such as LEO. The rapid growth of mega-constellations further complicates this process, which further emphasises the need for autonomous and AI-driven collision avoidance systems,
- Orbital slot allocation is particularly important for geostationary satellites that require fixed positions relative to Earth. The International Telecommunication Union (ITU) is responsible for coordinating these slots to prevent interference and overcrowding. However, as demand for orbital slots increases, STM must address issues of equity and accessibility, particularly ensuring that emerging spacefaring nations have fair opportunities to access these critical resources (Arnas & Linares, 2023),
- The equitable allocation of orbital slots is a technical challenge yet also a geopolitical one (Ailor, 2020). As commercial actors expand their presence in space, the need for transparent and collaborative allocation mechanisms becomes increasingly urgent,
- The legal and regulatory framework underpinning STM is essential for embracing and promoting international cooperation and accountability. The Outer Space Treaty 1967 does not address modern challenges, such as liability for collisions,

the removal of debris, or the regulation of commercial space activities (Munters, 2019),
- Many countries and private companies are reluctant to share detailed information about their spacecraft due to concerns over national security and intellectual property. This secrecy complicates the development of accurate and comprehensive space object catalogues, increasing the risk of miscommunication and accidents (Weeden, 2012).

5.3 Emerging Solutions and Innovations in Space Traffic Management

In response to these challenges, several innovative solutions and technologies are being developed to enhance the effectiveness of STM. AI and machine learning have emerged as powerful tools for improving collision prediction and optimising traffic flow. These technologies can process vast amounts of tracking data to identify potential risks and recommend mitigation strategies in real time (Cónego et al., 2024). Debris management technologies such as ADR, robotic arms, nets, and harpoons, are being tested to capture and de-orbit defunct satellites and fragments. While many of these technologies are still at the early stages in terms of technology readiness levels, they hold the potential to significantly reduce the risk of collisions and create a cleaner orbital environment (Sorge et al., 2021).

Advancements in satellite design also play a crucial role in improving STM. Modern satellites are increasingly equipped with propulsion systems that enable precise manoeuvring and end-of-life disposal. Some designs incorporate modular components that can be replaced or upgraded, reducing the need for new launches and minimising debris generation (McClintock, 2023), even though concerns and challenges associated with modular systems still remain unanswered.

Space-based tracking systems represent a significant leap forward in monitoring capabilities. By deploying sensors in orbit, these systems provide continuous coverage and higher accuracy than ground-based counterparts. This capability is particularly important for monitoring objects in remote orbits and dense regions like low Earth orbit (LEO) (Bhattacharjee, 2024; Manfletti et al., 2023).

5.4 The Future of Space Traffic Management

As space activities continue to expand, the future of Space Traffic Management (STM) will likely involve the integration of advanced technologies, robust policies, and global governance structures. One of the most promising developments is the use of artificial intelligence (AI) to enhance collision prediction and optimise traffic flow, enabling the management of large-scale satellite constellations with minimal human intervention; Accordingly, a pertinent area as a future research and practice is Trusted Autonomous Satellite Operations (TASO). Efforts to promote responsible space practices, such as the adoption of debris mitigation guidelines and the use of sustainable satellite materials, are essential to preserving the orbital environment. International agreements and regulatory frameworks must evolve to address these issues, ensuring that space remains accessible and safe for all (Khan & Ntantis, 2024).

Global cooperation will be critical to effective STM in the coming decades. Through collaboration among nations, private entities, and international organisations, humanity can and should create a unified approach to managing space traffic. Such cooperation is critical for the success of STM and also for the broader goal of sustainable space exploration and utilisation (Weeden, 2012; Drishti IAS, 2024). Collaboration internationally necessitates a sound governance foundation, and practical legal frameworks. The following section succinctly outlines some of the relevant frameworks.

6 Space Governance and Legal Frameworks

Effective governance and robust legal structures are fundamental for managing the exploration and utilisation of outer space. The advent of new actors, particularly private companies and non-governmental entities, presents both opportunities and challenges to the existing governance frameworks. As Freeland (2023) asserts, the increasing complexity of space activities necessitates adaptive and inclusive legal mechanisms to prevent fragmentation and ensure sustainability. The evolution of treaties such as the OST demonstrates the international community's attempt to maintain a shared vision for space governance, though significant gaps remain (Byers & Boley, 2023).

The integration of environmental and economic considerations further complicates governance. Cockell (2024) highlights that balancing commercial development with preservation of extraterrestrial environments is a pressing concern, particularly with the growing interest in planetary parks. Additionally, Marino and Cheney (2023) note that the utilisation of outer space increasingly intersects with terrestrial issues, which require governance frameworks that address both local and global priorities.

Previously, space governance was predominantly state-driven, with activities led by a few spacefaring nations (Clewley, 2021). However, the entry of private enterprises has transformed the landscape. This new landscape entails new challenges in regulation and equity. The commercialisation of space, while expanding access, risks impairing disparities in resource allocation and technological benefits. As Shagun and Anwer (2024) emphasise, equitable frameworks are vital to ensure that space activities do not reinforce existing global inequalities. The following subsections outline a number of relevant international treaties and agreements.

6.1 International Treaties and Agreements

The foundational treaties governing outer space include the Outer Space Treaty 1967, the Rescue Agreement (1968) (UNOOSA, 1968), the Liability Convention (1972) (UNOOSA, 1972), and the Moon Agreement (1979) (UNOOSA, 1979). These treaties establish core principles such as non-appropriation, peaceful use, and the benefit of all humankind:

- The Outer Space Treaty (OST) remains central to space governance. It prohibits national appropriation of outer space and celestial bodies, emphasising that space is 'the province of all humankind' (Clewley, 2021). However, its relevance is increasingly scrutinised as private actors take a prominent role in space activities. Scholars argue that the OST's general principles require more specific provisions to address issues such as resource extraction and space debris (Byers & Boley, 2023),
- The Rescue Agreement (1968) complements the OST by ensuring international cooperation in the return of astronauts and space objects. This treaty is instrumental in ensuring that there are constructive collaborative efforts between spacefaring nations, particularly during emergencies in space missions (Freeland, 2023). However, its applicability to private actors remains ambiguous, necessitating further clarity in light of growing commercial activities,
- The Liability Convention (1972) provides a framework for addressing damages caused by space objects. It establishes absolute liability for damage caused on Earth and fault-based liability for damages in outer space. This convention emphasises the principle of accountability, however it has been criticised for its limited applicability in addressing the complexities of modern space activities, including orbital collisions and space debris (Marino & Cheney, 2023),
- The Moon Agreement (1979), though less widely ratified, seeks to regulate activities on the Moon and other celestial bodies. It emphasises the use of such bodies for peaceful purposes and advocates for equitable sharing of benefits derived from space resources (Shagun & Anwer, 2024). Despite its noble intentions, major spacefaring nations have refrained from ratifying this treaty, citing concerns over its potential to stifle innovation and commercial interests (Cockell, 2024).

The development of international treaties has been further supported by resolutions and guidelines established under the auspices of the United Nations. For instance, the Guidelines for the Long-term Sustainability of Outer Space Activities (UNOOSA, 2018) provide non-binding recommendations to enhance safety and sustainability in space operations. Although these guidelines represent a significant step forward, their voluntary nature limits their enforceability, highlighting the need for stronger legal mechanisms (Byrd, 2021).

6.2 *Emerging Challenges in Governance*

6.2.1 The Rise of Private Actors

The increasing involvement of private companies are driven by profit motives, which sometimes conflict with the equitable and peaceful principles established by international treaties. As noted by Freeland (2023), the commercialisation of space necessitates a balanced approach to ensure safety and security without stifling innovation. Private companies have introduced a competitive dynamic into the space sector, which has traditionally been dominated by national space agencies. According to Marino and Cheney (2023), this shift has resulted in a fragmented governance landscape, as private actors often operate within regulatory gaps that were not anticipated when treaties like the OST were drafted. For instance, while the OST emphasises that outer space should benefit all humankind, it does not explicitly regulate the commercial exploitation of resources or establish mechanisms to enforce equitable benefit sharing. Moreover, the rapid pace of technological advancements in the private sector often outpaces the ability of international governance frameworks to adapt. Shagun and Anwer (2024) argue that private companies are pioneering technologies such as reusable launch vehicles and satellite mega-constellations, which have significant implications for orbital congestion and sustainability. These advancements, and the profit-driven nature of the efforts of the private entities highlight the urgent need for updated legal frameworks that can address emerging issues without stifling innovation.

The involvement of private actors has also sparked debates about liability and accountability. As Byrd (2021) notes, the Liability Convention (1972) places responsibility on launching states, however, the rise of private companies operating across multiple jurisdictions complicates the enforcement of these provisions. Questions arise about how liability is allocated when private entities cause harm, such as satellite collisions or environmental damage in space.

Soft law approaches, such as the Artemis Accords (NASA, 2020), have attempted to address these challenges by providing voluntary guidelines for responsible behaviour in space (Byrd, 2021). However, the reliance on non-binding agreements raises concerns about their effectiveness in ensuring compliance. Freeland (2023) emphasises that without binding international agreements, there is a risk that the activities of private actors could undermine the principles of peaceful use and equitable access enshrined in the OST. In addition to this, private companies often prioritise their commercial interests, which can lead to conflicts with broader societal goals such as environmental protection and equitable access. Cockell (2024) highlights that the exploitation of space resources by private entities could exacerbate inequalities between developed and developing nations unless robust governance mechanisms are established. For instance, the lack of clear legal frameworks for resource extraction could result in a 'first come, first served' approach that disproportionately benefits technologically advanced nations and their private sectors.

6.2.2 Space Debris Governance

The United Nations Office for Outer Space Affairs (UNOOSA) plays a crucial role in establishing guidelines to mitigate space debris. However, regulatory compliance remains uneven, particularly among private sector entities. Nozawa et al. (2023) argue for stronger enforcement mechanisms and international collaboration to ensure adherence to debris mitigation protocols. Moreover, institutional measures, such as taxation and licensing requirements, have been proposed to incentivise responsible behaviour among satellite operators. International coordination is further emphasised by the Inter-Agency Space Debris Coordination Committee (IADC), which provides recommendations to prevent the proliferation of debris. These include guidelines for the end-of-life disposal of satellites and rocket stages within a 25-year period (Hakima and Emami, 2018). Despite these efforts, gaps in implementation persist, as many nations and private companies lack the resources or incentives to comply fully with these measures (Forshaw et al., 2016).

The Outer Space Treaty (OST) of 1967 (UNOOSA, 1967) provides a foundational framework; however, it lacks enforceable provisions for debris mitigation, leaving the issue largely to national policies and voluntary guidelines, such as those established by the Inter-Agency Space Debris Coordination Committee (IADC). Recent studies suggest that updating this treaty to include specific obligations for debris removal and monitoring could significantly enhance global efforts (Svotina & Cherkasova, 2023). Additionally, bilateral, and multilateral agreements among leading spacefaring nations may serve as interim measures to address urgent concerns.

Technological advancements also play a role in supporting regulatory compliance. The development of Space Situational Awareness (SSA) systems enables more accurate tracking of debris, thus improving collision avoidance strategies and enhancing transparency among space operators (Shan et al., 2020). Furthermore, private-public partnerships (PPP) could facilitate the development and deployment of ADR technologies, which ultimately aligns commercial interests with international safety objectives (Mark & Kamath, 2019). That said, without coordinated international action, the escalating costs associated with debris-related damages and mitigation efforts could deter investment in space activities, which can ultimately undermine the long-term viability of this critical sector (Nozawa et al., 2023).

One innovative proposal is the establishment of international standards for debris mitigation. Cockell (2024) argues that "adopting universally recognised debris mitigation protocols will play a crucial role in safeguarding orbital environments for future generations". Such standards could include mandatory end-of-life disposal plans for satellites and incentives for designing satellites with reusability and recyclability in mind. Moreover, the concept of a 'polluter pays' principle has been suggested as a means to encourage compliance with debris mitigation measures. This approach, as detailed by Shagun and Anwer (2024), could involve levying fines or penalties on entities that fail to adhere to agreed-upon guidelines, hence internalising the environmental costs of space activities. Although this principle has gained traction in terrestrial environmental governance, its implementation in space law remains a contentious issue.

The concept of planetary parks, akin to national parks on Earth, offers a potential framework for balancing preservation and development. Cockell (2024) suggests that such parks could safeguard scientifically and culturally significant sites while allowing commercial activities in designated areas. In addition, the development of space traffic management systems, as proposed by Freeland (2023), could ensure the safe and equitable use of orbital pathways, mitigating risks associated with debris and overcrowding.

6.2.3 Equity and Benefit Sharing

Shagun and Anwer (2024) emphasise the need for equitable frameworks to ensure that emerging space-faring nations and underrepresented groups have access to the benefits of space activities. The growing disparity in space activity benefits highlights the need for more inclusive governance structures. Freeland (2023) argues that the dominance of technologically advanced nations often marginalises less-developed countries in decision-making processes. This imbalance is particularly evident in the allocation of geostationary orbital slots and radio frequencies, which are finite resources critical for satellite operations. Current frameworks lack mechanisms to ensure equitable access, perpetuating inequalities between space-faring and non-space-faring nations. Shagun and Anwer (2024) further contend that initiatives such as space-based solar power (SBSP) could provide significant benefits to underprivileged regions by addressing energy disparities. However, without robust international agreements to regulate equitable access and distribution, such projects risk exacerbating existing inequalities. For instance, high-income countries might monopolise these technologies, sidelining low- and medium-income nations despite their pressing energy needs.

The concept of benefit-sharing also extends to scientific advancements. Byers and Boley (2023) note that research conducted in microgravity environments, such as on the International Space Station, often yields significant medical and technological innovations. However, the dissemination of these benefits is often restricted to entities from high-income countries. Establishing mandatory benefit-sharing agreements could help ensure that advancements derived from space research contribute to global development. Accordingly, Freeland (2023) suggests that incorporating equity-focused clauses into future treaties or amending existing ones could provide a pathway to more inclusive governance. Additionally, capacity-building initiatives, such as technology transfer and capacity building programmes, could empower emerging space nations to participate more effectively in the global space economy.

6.3 The Role of Soft Law

Soft law mechanisms, such as guidelines and non-binding agreements, have gained prominence in addressing the limitations of hard law. These include the United Nations (UN) Guidelines for the Long-term Sustainability of Outer Space Activities and the Artemis Accords. While soft law is flexible and adaptive, its non-binding nature limits enforceability, potentially undermining efforts to address global challenges like space debris and equitable resource distribution (Freeland, 2023). The Artemis Accords emphasise peaceful exploration and resource-sharing principles (Byrd, 2021). These accords have facilitated dialogue but remain contentious due to their lack of universal adoption and perceived alignment with certain national interests (Freeland, 2023). Moreover, the reliance on soft law highlights gaps in binding international treaties. The absence of enforceable mechanisms to regulate commercial activities and ensure compliance with sustainability measures highlights the need for a more structured approach (Cockell, 2024). This is particularly relevant in mitigating the risks posed by mega-constellations, as voluntary adherence to debris guidelines often falls short of addressing the scale of the problem (Marino & Cheney, 2023).

Critics argue that although soft law encourages innovation, it risks creating a fragmented legal landscape that favours dominant space-faring nations (Shagun & Anwer, 2024). To address this, there is a growing call for hybrid frameworks that combine the flexibility of soft law with the enforceability of hard law provisions. Such an approach could provide a balanced pathway for regulating emerging challenges while ensuring inclusivity and equity in space governance (Freeland, 2023).

7 Conclusion

Sustainability in space is affected by a number of pressing challenges. Space debris management is a global imperative requiring immediate technical, regulatory, and economic measures. With the escalating risks of the Kessler Syndrome and the increasing reliance on satellite infrastructure, proactive ADR technologies and robust international policies are essential to ensure the sustainability of space activities for future generations. Sound space traffic management technologies and policies are also needed to ensure equity in access, and also prevent collisions and catastrophes that can result in problems such as space debris and environmental pollutions. The environmental impacts caused by space launches also affect flora and fauna, both land and water and the also causes air pollution, necessitating advances in cleaner technologies for the space launches and operations.

Looking forward, effective governance and legal structures will play a crucial role in addressing the complexities of space debris management. The next section explores the foundational treaties and agreements, such as the Outer Space Treaty, which govern the use of space. It also examines the challenges of enforcing these frameworks, particularly with the advent of new actors in space exploration, including private companies and non-governmental entities. Understanding these legal and regulatory landscapes will be pivotal in creating a sustainable and cooperative approach to space governance.

References

Ailor, W. (2020). Evolution of space traffic and space traffic management. In *Handbook of space security* (pp. 299–317). Springer International Publishing. https://doi.org/10.1007/978-3-030-23210-8_88

Arnas, D., & Linares, R. (2023). Non-self-intersecting trajectories and their applications to satellite constellation design and orbital capacity. *Journal of Guidance, Control, and Dynamics, 46*(3), 417–430. https://doi.org/10.2514/1.G007208

Bhattacharjee, N. (2024). *Global push for cooperation as space traffic crowds Earth orbit, Reuters.* Retrieved January 2, 2025, from https://www.reuters.com/science/global-push-cooperation-space-traffic-crowds-earth-orbit-2024-12-02/

Brown, T. F. M., Bannister, M. T., & Revell, L. E. (2024). Envisioning a sustainable future for space launches: A review of current research and policy. *Journal of the Royal Society of New Zealand, 54*(3), 273–289. https://doi.org/10.1080/03036758.2022.2152467

Bushnell, D., & Moses, R. (2018). *Commercial space in the age of "new space", reusable rockets and the ongoing tech revolutions.* Hampton, VA 23681-2199. Retrieved January 2, 2025, from https://ntrs.nasa.gov/citations/20180008444

Byers, M., & Boley, A. (2023). *Who owns outer space?* Cambridge University Press. https://doi.org/10.1017/9781108597135

Byrd, L. C. (2021). Soft law in space: A legal framework for extraterrestrial mining. *Emory LJ, 71*, 801.

Chrysaki, M. (2020). The sustainable commercialisation of space: The case for a voluntary code of conduct for the space industry. *Space Policy, 52*, 101375. https://doi.org/10.1016/j.spacepol.2020.101375

Clewley, P. T. (2021). Newspace: The rise of the private space industry is threatening the current legal framework governing outer space. *J. High Tech. l., 21*, 354.

Cockell, C. S. (2024). Planetary parks twenty years on. Balancing space protection with development. *Space Policy, 70*, 101647. https://doi.org/10.1016/j.spacepol.2024.101647

Cónego, L., Pinto, R., & Gonçalves, G. (2024). Digital transformation in manufacturing: The synergy of chatbots and tailored gamification strategies. *Procedia Computer Science, 237*, 171–178. https://doi.org/10.1016/j.procs.2024.05.093

Dallas, J. A., et al. (2020). The environmental impact of emissions from space launches: A comprehensive review. *Journal of Cleaner Production, 255*, 120209. https://doi.org/10.1016/j.jclepro.2020.120209

Drishti IAS. (2024). *Global cooperation in space debris management, Drishti IAS.* Retrieved January 2, 2025, from https://www.drishtiias.com/pdf/1733341317.pdf

Forganni, A. (2023). Legal considerations on space commercialisation. In *The commercialisation of space* (pp. 211–227). Routledge.

Forshaw, J. L., et al. (2016). RemoveDEBRIS: An in-orbit active debris removal demonstration mission. *Acta Astronautica, 127*, 448–463. https://doi.org/10.1016/j.actaastro.2016.06.018

Freeland, S. (2023). Promoting safety and security: Challenges for the future governance of space activities. *Advanced Space Law, 12*. https://doi.org/10.29202/asl/12/8

Gharehbaghi, K. (2017). Modular high-rise construction: An alternative building system. *The International Journal of the Constructed Environment, 8*(3), 15–25. https://doi.org/10.18848/2154-8587/CGP/v08i03/15-25

Görner, M., et al. (2024). The DLR moon-mars test site for robotic planetary exploration. In *2024 International Conference on Space Robotics (iSpaRo)* (pp. 245–252). IEEE. https://doi.org/10.1109/iSpaRo60631.2024.10687411
Gottlieb, J. (2019). Space colonization and existential risk. *Journal of the American Philosophical Association, 5*(3), 306–320. https://doi.org/10.1017/apa.2019.12
Hakima, H., & Emami, M. R. (2018). Assessment of active methods for removal of LEO debris. *Acta Astronautica, 144*, 225–243. https://doi.org/10.1016/j.actaastro.2017.12.036
Hampson, J. (2017). The future of space commercialization. *Niskanen Center Research Paper, 25.*
Hilton, S., et al. (2019). Space traffic management: Towards safe and unsegregated space transport operations. *Progress in Aerospace Sciences, 105*, 98–125. https://doi.org/10.1016/j.paerosci.2018.10.006
Kessler, D. J., & Cour-Palais, B. G. (1978). Collision frequency of artificial satellites: The creation of a debris belt. *Journal of Geophysical Research: Space Physics, 83*(A6), 2637–2646. https://doi.org/10.1029/JA083iA06p02637
Khan, M. V. H., & Ntantis, E. L. (2024). Space debris: Overview and mitigation strategies. In *Proceedings of the 8th International Conference on Research, Technology and Education of Space, H-Space* (pp. 25–26).
Koroleva, T. V., et al. (2018). The environmental impact of space transport. *Transportation Research Part D: Transport and Environment, 58*, 54–69. https://doi.org/10.1016/j.trd.2017.10.013
Lonsdale, J.-A., & Phillips, C. (2021). Space launches and the UK marine environment. *Marine Policy, 129*, 104479. https://doi.org/10.1016/j.marpol.2021.104479
Lubert, C. P. (2018). *From Sputnik to SpaceX®: 60 Years of rocket launch acoustics, acoustics today*. Retrieved January 2, 2025, from https://acousticstoday.org/from-sputnik-to-spacex-60-years-of-rocket-launch-acoustics-caroline-p-lubert/
Maclay, T., & McKnight, D. (2021). Space environment management: Framing the objective and setting priorities for controlling orbital debris risk. *Journal of Space Safety Engineering, 8*(1), 93–97. https://doi.org/10.1016/j.jsse.2020.11.002
Manfletti, C., Guimarães, M., & Soares, C. (2023). AI for space traffic management. *Journal of Space Safety Engineering, 10*(4), 495–504. https://doi.org/10.1016/j.jsse.2023.08.007
Maraš, D., & Dangubić, M. (2022). Cooperation between government agencies and private companies in space: The case of the United States. *Astropolitics, 20*(2–3), 226–237. https://doi.org/10.1080/14777622.2022.2146495
Marino, A., & Cheney, T. (2023). Centring environmentalism in space governance: Interrogating dominance and authority through a critical legal geography of outer space. *Space Policy, 63*, 101521. https://doi.org/10.1016/j.spacepol.2022.101521
Mark, C. P., & Kamath, S. (2019). Review of active space debris removal methods. *Space Policy, 47*, 194–206. https://doi.org/10.1016/j.spacepol.2018.12.005
McClintock, B. (2023). *International space traffic management charting a course for long-term sustainability*. Retrieved January 2, 2025, from https://apps.dtic.mil/sti/citations/trecms/AD1202870
Mitchell, R. (2018). Into the final frontier: The expanse of space commercialization. *Mo. l. Rev., 83*, 429.
Muelhaupt, T. J., et al. (2019). Space traffic management in the new space era. *Journal of Space Safety Engineering, 6*(2), 80–87. https://doi.org/10.1016/j.jsse.2019.05.007
Munters, W. (2019). Active debris removal, international environmental law, and the collective management of risk: Foundations of an international system for space traffic management. In *Space security and legal aspects of active debris removal* (pp. 131–154). https://doi.org/10.1007/978-3-319-90338-5_8
Musk, E. (2017). Making humans a multi-planetary species. *New Space, 5*(2), 46–61. https://doi.org/10.1089/space.2017.29009.emu
NASA. (2020). *Artemis Accords, Artemis Accords*. Retrieved January 6, 2025, from https://www.nasa.gov/artemis-accords/

NASA. (2021). *NASA's management of Artemis mission, NASA*. Retrieved January 2, 2025, from https://oig.nasa.gov/docs/IG-22-003.pdf

Naser, M. Z., & Chehab, A. I. (2018). Materials and design concepts for space-resilient structures. *Progress in Aerospace Sciences, 98*, 74–90. https://doi.org/10.1016/j.paerosci.2018.03.004

Naser, M. Z., & Chen, Q. (2021). Extraterrestrial construction in Lunar and Martian environments. In *Earth and Space 2021* (pp. 1200–1207). American Society of Civil Engineers. https://doi.org/10.1061/9780784483374.111

Nozawa, W., et al. (2023). To what extent will space debris impact the economy?. *Space Policy, 66*, 101580. https://doi.org/10.1016/j.spacepol.2023.101580

Pinto, F., et al. (2020). Towards automated satellite conjunction management with Bayesian deep learning. *ArXiv*, abs/2012.12450. https://api.semanticscholar.org/CorpusID:229363495

Shagun, S., & Anwer, N. (2024). Space-based solar power: Legal frameworks and sustainable development perspectives. *Discover Applied Sciences, 6*(12), 637. https://doi.org/10.1007/s42452-024-06259-5

Shan, M., Guo, J., & Gill, E. (2020). An analysis of the flexibility modeling of a net for space debris removal. *Advances in Space Research, 65*(3), 1083–1094. https://doi.org/10.1016/j.asr.2019.10.041

Sharapova, A. V., et al. (2020). Snow pollution by nitrogen-containing substances as a consequence of rocket launches from the Baikonur Cosmodrome. *Science of the Total Environment, 709*, 136072. https://doi.org/10.1016/j.scitotenv.2019.136072

Simpson, M. (2014). Future of space commercialization–Mining asteroids and celestial bodies. *Commercialization on Space: Opportunities and Challenges*, 176–183.

Sorge, M. E., Ailor, W. H., & Muelhaupt, T. J. (2021). *Space traffic management: The challenges of large constellations and orbital debris, and rapid changes in space operations, centre for space policy and strategy*. Retrieved January 2, 2025, from https://aerospace.org/sites/default/files/2020-09/Sorge_STM_20200915.pdf

Spudis, P. D. (2016). *The value of the Moon: How to explore, live, and prosper in space using the Moon's resources*. Smithsonian Institution.

Steffen, O. (2022). Explore to exploit: A data-centred approach to space mining regulation. *Space Policy, 59*, 101459. https://doi.org/10.1016/j.spacepol.2021.101459

Sturdevant, R. W. (2024). The evolution of commercial space and its integration with US military space. In *The rise of the commercial space industry* (pp. 253–290). Springer International Publishing. https://doi.org/10.1007/978-3-031-63410-9_13

Svotina, V. V., & Cherkasova, M. V. (2023). Space debris removal – Review of technologies and techniques. Flexible or virtual connection between space debris and service spacecraft. *Acta Astronautica, 204*, 840–853. https://doi.org/10.1016/j.actaastro.2022.09.027

University of Surrey. (2019). *Harpoon successfully captures space debris*. University of Surrey. Retrieved January 2, 2025, from https://www.surrey.ac.uk/news/harpoon-successfully-captures-space-debris

UNOOSA. (1967). *2222 (XXI). Treaty on principles governing the activities of states in the exploration and use of outer space, including the Moon and other celestial bodies, Outer Space Treaty (OST)*. Retrieved January 6, 2025, from https://www.unoosa.org/oosa/en/ourwork/spacelaw/treaties/outerspacetreaty.html

UNOOSA. (1968). *Agreement on the Rescue of Astronauts, the Return of Astronauts and the Return of Objects Launched into Outer Space* [WWW Document]. *Rescue Agreement*. https://www.unoosa.org/oosa/en/ourwork/spacelaw/treaties/introrescueagreement.html. Accessed January 6, 2025.

UNOOSA. (1972). *Convention on international liability for damage caused by space objects* [WWW Document]. *liability convention*. https://www.unoosa.org/oosa/en/ourwork/spacelaw/treaties/introliability-convention.html. Accessed January 6, 2025.

UNOOSA. (1979). *Agreement governing the activities of states on the Moon and other celestial bodies* [WWW Document]. *Moon agreement*. https://www.unoosa.org/oosa/en/ourwork/spacelaw/treaties/intromoon-agreement.html. Accessed January 6, 2025.

UNOOSA. (2018). Guidelines for the long-term sustainability of outer space activities. *Guidelines for the long-term sustainability of outer space activities* [Preprint]. UNOOSA. Retrieved January 6, 2025, from https://www.unoosa.org/res/oosadoc/data/documents/2018/aac_1052018crp/aac_1052018crp_20_0_html/AC105_2018_CRP20E.pdf

Van Camp, C., & Peeters, W. (2022). A world without satellite data as a result of a global cyber-attack. *Space Policy*, *59*, 101458. https://doi.org/10.1016/j.spacepol.2021.101458

Weeden, B. (2012). The economics of space sustainability. *The Space Review*, *4*. Retrieved January 2, 2025, from https://www.thespacereview.com/article/2093/1#:~:text=by%20Brian%20Weeden,Monday%2C%20June%204%2C%202012&text=A%20crucial%20element%20of%20space,it%20poses%20to%20active%20satellites

Wu, C., et al. (2024). A comprehensive review of extraterrestrial construction, from space concrete materials to habitat structures. *Engineering Structures*, *318*, 118723. https://doi.org/10.1016/j.engstruct.2024.118723

Zhao, Y., Zhang, F., & Huang, P. (2020). Capture dynamics and control of tethered space net robot for space debris capturing in unideal capture case. *Journal of the Franklin Institute*, *357*(17), 12019–12036. https://doi.org/10.1016/j.jfranklin.2020.04.037

Zubrin, R. (2019). Why we earthlings should colonize mars!. *Theology and Science*, *17*(3), 305–316. https://doi.org/10.1080/14746700.2019.1632519

Zuniga, A. F., et al. (2015). Lunar COTS: An economical and sustainable approach to reaching mars. In *AIAA SPACE 2015 Conference and Exposition*. American Institute of Aeronautics and Astronautics. https://doi.org/10.2514/6.2015-4408

Planetary Penetrators for Space Mining

Viduranga Shenal Landers

Abstract Space mining represents the next frontier for resource acquisition. As humanity sets its sights on the vast resources of our solar system, the demand for efficient and cost-effective tools grows. This chapter explores the potential of planetary penetrators as such a tool for identifying resources and selecting mining sites on celestial bodies. These high-velocity penetrators can be deployed in swarms to rapidly assess subsurface compositions across extensive areas. Equipped with miniaturized sensors, penetrators analyze regolith composition and gather crucial data on the presence of valuable resources like water ice, rare earth elements, and precious metals. The chapter examines technological advancements, introduces and compares various penetrator mission concepts, and highlights penetrator advantages over traditional methods through direct, in-situ measurements for precise resource mapping at lower cost. Simulation-based analysis is presented to validate the performance of novel concepts. Finally, the chapter addresses legal and ethical considerations surrounding the use of penetrators, advocating for responsible design and deployment strategies illustrating how penetrators can be instrumental in facilitating sustainable space resource utilization.

Keywords Space mining · Planetary penetrators · Resource prospecting · In-situ resource utilization

1 Introduction

The exploration and utilization of extraterrestrial resources, often termed space mining, is rapidly transitioning from science fiction to a tangible prospect pursued by both scientific communities and commercial entities. As the global population grows and technology advances, humanity's footprint expands, placing mounting pressure on terrestrial resources due to increasing demand and environmental concerns. The seemingly boundless resources available on the Moon, Mars, asteroids, and

V. S. Landers (✉)
University of Colombo School of Computing, Colombo, Sri Lanka
e-mail: vidurangalanders@gmail.com

H. Jahankhani et al. (eds.), *Space Mining*, Space Law and Policy,
https://doi.org/10.1007/978-981-95-5543-7_8

other celestial bodies present a compelling alternative (Noorda et al., 2024). These extraterrestrial bodies contain vast quantities of valuable materials, including water ice, industrial metals, precious metals, and other volatiles. These could potentially support Earth's economy while also providing the foundational materials necessary for building a sustainable infrastructure for deeper space exploration and colonization (Bushnell, 2021).

However, realizing the potential of space resources involves overcoming significant technical, economic, and logistical hurdles. Identifying minable deposits and accurately characterizing their composition, concentration, and depth are critical prerequisites for any mining effort (Sanders, 2018). Traditional space exploration methods, often involving complex landers or rovers, are inherently expensive. Such missions frequently cost billions of dollars, and typically provide detailed information about only a very limited geographic area (McCarthy, 2021). While remote sensing from orbit offers broader coverage, it often lacks the necessary resolution or subsurface depth to definitively assess resource potential, particularly for deposits buried beneath the surface (Oleg et al., 2021).

Planetary penetrators emerge as a complementary and potentially transformative technology designed to address these limitations. A planetary penetrator is a rigid probe engineered to survive a high-velocity impact on the surface of a celestial body, using its kinetic energy to bury itself and directly access and analyze subsurface materials (Lorenz, 2011). By eliminating the need for complex and heavy soft-landing systems, penetrators significantly reduce mission cost and complexity for subsurface exploration. Furthermore, their relatively lower mass and cost allow for mission architectures involving the deployment of multiple probes (swarms) across a wide area, facilitating distributed, in-situ sampling and prospecting experiments.

This chapter explores the potential of planetary penetrators as critical tools for space resource identification and eventual mining. Section 2 outlines the emerging space economy and the necessity for space resource identification, covering both remote sensing and in situ prospecting methods. Section 3 traces the historical development of penetrators, highlights past missions, and discusses key technological advancements enabling modern penetrator designs. Section 4 presents various novel mission concepts and advanced techniques that could define the next generation of penetrator applications. Section 5 assesses the viability of several penetrator concepts using numerical simulations of surface penetration. Crucial legal, ethical, and sustainability considerations are addressed in Sect. 6. Finally, Sect. 7 concludes the chapter with a forward-looking perspective on the role penetrators might play in the future sustainable space mining ecosystem.

In essence, by offering potentially low-cost and widespread access to the subsurface, planetary penetrators have the potential to serve as a key enabler for space mining, bridging the critical gap between orbital reconnaissance and effective resource extraction in the solar system.

2 The Space Economy and Space Mining

Space mining is formally defined as the in-situ and ex-situ utilization of space resources (Noorda et al., 2024). Examples of in-situ resource utilization (ISRU) include using water ice for propellants and employing other resources for in-space manufacturing and infrastructure development. Ex-situ utilization involves sending extracted resources back to the Earth or, eventually, to future space colonies (Hepp et al., 2014). Although still in its early stages, space mining has attracted significant interest and investment over the past decade, driven by the converging technological advancements, growing terrestrial resource constraints, and ambitious space exploration goals. This section explores the context of the expanding space economy, outlines current initiatives in space mining, and discusses the projected future of this potentially transformative industry.

2.1 The Emerging Space Economy

The twenty-first century has witnessed a dramatic evolution of the space sector, often termed the "new space economy". The global space industry is projected to experience exponential growth, fueled by innovations like reusable launch vehicles that drastically lower access costs, increased private investment, and expanding commercial applications beyond traditional satellites. Analysts from institutions like European Space Agency (ESA) and World Economic Forum (WEF) estimate the space market's value between $530 and $630 billion in 2024, forecasting it to surpass $1.8 trillion by 2035 with an annual growth rate of 9% (ESA, 2025; WEF, 2024). Within this dynamic landscape, space mining represents a high-potential market segment. Market analysts from Emergen Research valued the space mining market size at $1.22 billion in 2021 and anticipate the industry reaching approximately $8 billion annually by the early 2030s (Emergen Research, 2022).

The rationale for space mining stems from two primary drivers. Firstly, the increasing demand for critical minerals and rare metals (such as platinum-group metals and rare earth elements essential for modern technologies and clean energy transitions) coupled with the finite nature and environmental impact of terrestrial extraction, makes asteroids and celestial bodies attractive alternative sources (Driessen & Henckens, 2016). Secondly, the strategic importance of ISRU is increasingly recognized as critical for enabling sustainable, long-term human exploration (Bushnell, 2021). Resources like water ice, which are abundant in lunar polar craters and certain asteroids, can be processed into rocket propellant or used for life support. Government agencies like National Aeronautics and Space Administration (NASA), through its Artemis program aimed at establishing a sustainable lunar presence, explicitly prioritize ISRU development. They view local resource utilization as key

Table 1 Solar system resources—excerpt from (Mueller & Susante, 2012)

	Resources				
	Moon	Mars	Asteroids	Comets	Outer planets
Regolith	H_2O ice, O_2, OH, granular powder, shielding, insulation, structural, manufacturing feedstock	H_2O ice, Rocky with sand, shielding, structural, insulation, manufacturing, feedstock	Varies—granular powder, TBD	H_2O ice TBD	TBD
Volatiles	H_2O, CO, H_2, Hg, H_2S, NH_3, He_3, CH_4, Ca, C_2H_6, CH_3OH, C_2H_4	CO_2, O, N_2, H_2O, Cl, Br, Fl, CO, TBD	H_20, TBD	H_20, CH_4, TBD	H, He, CH_4, N_2, H_2O, CO_2, CO, NH_4, C_2H_6 TBD
Minerals	Feldspars, pyroxines, olivines, oxides	Basalt, andecite, hematite, silicates, sulfates, carbonates, pegmatities, oxides, TBD	Silicates, oxides, TBD	TBD	Silicates, Oxides, TBD
Metals	Al, Fe, Ti, Mg, Ca, TBD	Al, Fe, Ti, Mg, Ca, Na, Cu, Zn, Pb, Cr, Ni, Co, PGM, TBD	Fe, Ni, Ti, PGM etc. TBD	TBD	Fe, TBD
Rare Earth elements	Yes, associated with KREEP	TBD	TBD	TBD	TBD
Water	Limited—poles	Yes	Yes	Yes	Abundant

to establishing permanent bases on the Moon and paving the way for Mars exploration (Sanders & Leinhenz, 2024). A summary of the key resources available on various solar system bodies is provided in Table 1.

2.2 *Space Mining*

Space mining holds the potential to revolutionize resource availability, enable a truly sustainable presence in space, and contribute significantly to the future space economy. Currently, the state of space mining is primarily characterized by prospecting, technology development, and demonstration missions rather than operational extraction. Both private companies and governmental space agencies are actively laying the groundwork while navigating challenges such as mitigating the risks posed by orbital debris, ensuring the long-term sustainability of operations, and navigating the evolving legal and regulatory landscape.

A wave of private startups is pursuing various approaches. Companies like Astro-Forge are targeting asteroid mining, specifically for precious metals, and have launched initial test payloads with plans for asteroid rendezvous missions (Aubert, 2025). Others, such as the Japanese firm ispace, focus on lunar resources, developing landers and rovers aimed at locating and eventually harvesting lunar water ice, despite early mission setbacks highlighting the technical difficulties (ispace, 2024). Firms like Astrobotic, Intuitive Machines, and Firefly are among the 14 providers developing lunar landers and rovers under NASA's Commercial Lunar Payload Services (CLPS) program, which will carry payloads designed to detect and analyze lunar resources (NASA, 2025). TransAstra is developing novel concepts like optical mining using concentrated sunlight to extract volatiles from asteroids (David, 2015). While pioneering companies like Planetary Resources faced challenges and were eventually acquired, they catalyzed initial interest and technological progress in the field (Boyle, 2020).

Governmental initiatives are equally crucial. Sample return missions, such as JAXA's Hayabusa2 from asteroid Ryugu (JAXA, 2018) and NASA's OSIRIS-REx from asteroid Bennu (Lauretta et al., 2017), along with China's Chang'e 5 lunar sample return (Chen et al., 2023), have provided invaluable ground truth about the composition and structure of these bodies, directly informing resource potential assessments. NASA's canceled VIPER rover mission aimed to map water ice deposits at the Moon's south pole, a critical step for ISRU (Doyle, 2024). Furthermore, technology demonstrations like the MOXIE experiment on the Mars Perseverance rover, which is capable of producing oxygen from Martian CO_2, prove the feasibility of key ISRU processes (Hecht et al., 2021). International collaboration is also evident, with entities like ESA and the Luxembourg Space Agency establishing the European Space Resources Innovation Centre (ESRIC) to fund research and development (Luxembourg Space Agency, 2025).

Collectively, these public and private efforts are focused on identifying resource locations, characterizing their composition and concentration, and testing the technologies required for extraction, processing, and utilization in the challenging space environment.

2.3 *Resource Identification: The Foundation of Space Mining*

Successful space mining hinges on the initial discovery and identification of viable resource deposits. Unlike terrestrial mining exploration, which benefits from direct access, space resource prospecting faces the challenges of scouting potential targets across vast distances and locating valuable materials in sufficient concentrations

(Sanders, 2018). Achieving this requires a sophisticated strategy that typically combines broad remote reconnaissance with targeted ground verification.

Initial exploration relies heavily on remote sensing technologies deployed via spacecraft or Earth-based telescopes, gathering data about a celestial body's composition and surface properties without physical contact. Several key techniques, building upon methods used in numerous space exploration missions, are applicable. The most straightforward method is high-resolution optical imaging. This offers visual context on surface features, identifies potential resource traps like shadowed craters, and aids mission planning. An advancement is multispectral or hyperspectral imaging, analyzing the spectrum of electromagnetic radiation reflected or emitted by the body to identify the unique chemical fingerprints of minerals, elements, and compounds, similar to spectroscopy (Qian, 2023). Various forms of spectroscopy itself—including Visible/Near-Infrared, Thermal Infrared, and Gamma-Ray/Neutron analysis—further probe chemical composition by examining specific radiation signatures. For example, detecting hydrogen via neutron spectroscopy serves as a popular indicator for water ice presence (Lawrence et al., 2012). Radar and microwave sensing not only map surface topology but can also penetrate the ground to identify buried structures or resources and differentiate materials like metals based on reflectivity (Ulaby et al., 2014). Additionally, gravitational and magnetic mapping offer insights into internal structure, density variations, and concentrations of magnetic minerals (Hinze et al., 2013).

Despite their power for initial surveys and identifying regions of interest, remote sensing methods have inherent limitations. Their penetration depth is often confined to the near-surface, potentially missing deeper deposits, and the spatial resolution for compositional mapping can be coarse, necessitating confirmation through direct measurement (Oleg et al., 2021).

Consequently, in-situ prospecting technologies, involving direct contact with the surface or subsurface, are essential for providing definitive ground truth and detailed characterization. These range from stationary landers equipped with instruments like spectrometers, drills, and microscopes for focused site analysis, to mobile rovers capable of surveying wider areas and sampling diverse locations with sophisticated analytical suites (Gellert & Clark, 2015). Accessing resources buried beneath the regolith invariably requires drilling or similar mechanical or thermal methods. Sample return missions offer the highest fidelity analysis by returning materials for study in advanced Earth-based laboratories. However, despite providing invaluable high-fidelity data, landers, rovers, and sample return missions are generally complex, expensive, and limited in their spatial coverage due to limitation to single sites, slow traverses, or pinpoint collection points.

This highlights a critical need for complementary, cost-effective methods capable of achieving more widespread subsurface sampling and ground-truthing across promising regions identified by remote sensing. Planetary penetrators, which use high-velocity impact to emplace instruments meters below the surface without complex drilling systems, offer a potential solution to bridge this gap. They complement orbital surveys and augment lander or rover missions by providing targeted subsurface data points across broader areas (Lorenz, 2011).

3 Planetary Penetrator Concepts and Missions

The concept of using high-velocity penetrators is not new. It has been employed for military purposes since World War II as "bunker buster" bombs (US Army, 1946). The use of planetary penetrators for space exploration has also been explored for decades (Lorenz, 2011). While no penetrator mission has yet fully succeeded in returning scientific data from beneath the surface of another celestial body, numerous projects—both flown and conceptual—have significantly advanced the technology and provided invaluable lessons. This section reviews the evolution of penetrator concepts, highlights key missions and their outcomes, and discusses the critical technological advancements that make modern penetrators increasingly viable tools, particularly for demanding applications like space mining.

3.1 Concept and Historical Development

A planetary penetrator is fundamentally a projectile engineered to impact a planetary surface at high velocity, embed itself to depths ranging from centimeters to several meters, and survive the immense deceleration forces. Its core components typically include a robust forebody to withstand impact, a protective housing for scientific instruments, a communication system for data transmission, and a power source like batteries or a radioisotope thermoelectric generator (RTG). The penetration depth is determined by the penetrator's mass, impact velocity, shape, and the target surface's mechanical properties (Lorenz, 2011).

Penetrators offer several distinct advantages. They provide subsurface access without complex drilling, allow for deployment at multiple sites from a single spacecraft, are relatively power-efficient during deployment, and can reach hazardous or remote locations inaccessible to traditional landers or rovers. Key scientific goals often include analyzing subsurface composition and geochemistry for resource identification, studying internal structure via seismology, measuring heat flow to understand thermal evolution, and searching for compounds like water ice or organics.

The idea emerged in the 1960s, leveraging military technology adapted for scientific use. NASA's Langley Research Center initiated studies in the 1960s, particularly for the Moon (McCarty et al., 1964). Throughout the 1970s and 1980s, NASA refined Mars penetrator designs linked to potential sample return missions (Hughes, 1974) as well as comet exploration penetrators (Adams et al., 1986), conducting impact tests and developing shock-resistant instrumentation.

The 1990s saw the first concrete mission attempts. Russia's Mars 96 mission included two sophisticated penetrator probes intended for Martian deployment, though the mission tragically failed during launch in 1996 (Space Research Institute, 2002). Concurrently, NASA developed the Deep Space 2 (DS-2) microprobes, which flew with the Mars Polar Lander in 1999, representing the first actual deployment

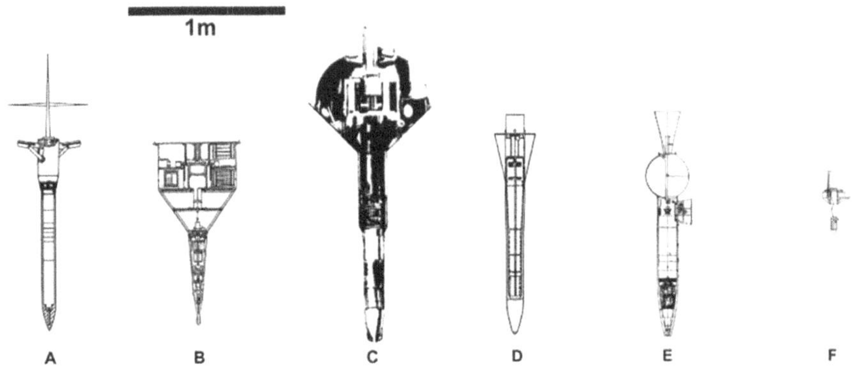

Fig. 1 Excerpt from Lorentz's paper on the origins and history of penetrators showing historic penetrators. **a** An ACUSID air-dropped acoustic monitor used in Vietnam Conflict. **b** Comet Rendezvous Asteroid Flyby (CRAF) penetrator. **c** the Mars-96 penetrator. **d** NASA proposed 1970s Mars penetrator. **e** Lunar-A penetrator. **f** DS-2 (Lorenz, 2011)

attempt on another planet (Smrekar et al., 1999). Several of these historical penetrator concepts are illustrated in Fig. 1 (Lorenz, 2011).

3.2 Significant Planetary Penetrator Missions

NASA's Deep Space 2 marked the pioneering effort to utilize penetrators extraterrestrially. Two small microprobes, each weighing approximately 2.4 kg, were to be released during the Mars Polar Lander's atmospheric entry. They were designed to strike the Martian south polar region at roughly 190 m/s and penetrate up to 0.6 m (Smrekar et al., 1999). Each featured a hardened forebody containing instruments that would bury itself, connected by a tether to an aftbody remaining on the surface to communicate via an antenna. Surviving the estimated 30,000 g impact necessitated custom high-g electronics and batteries, with components potted for cushioning. Unfortunately, communication was never established after arrival in December 1999. Post-mission analysis pointed to potential failures resulting from the extreme impact shock or a breakdown in the communication link. Despite its failure to return data, DS2 demonstrated the feasibility of high-velocity delivery and highlighted critical areas for improvement such as enhanced impact survivability and robust subsurface communication strategies.

Around the same period, Russia's ambitious Mars' 96 mission included two much larger penetrators, each around 45 kg penetrators. These were designed to carry a comprehensive instrument suite—including cameras, meteorological sensors, seismometer, and spectrometers—and penetrate 5–6 m into the Martian soil, operating for up to a year. Significantly, they planned a multi-stage descent using aerobraking and inflatable decelerators to reduce impact velocity to 80 m/s. This comparatively soft

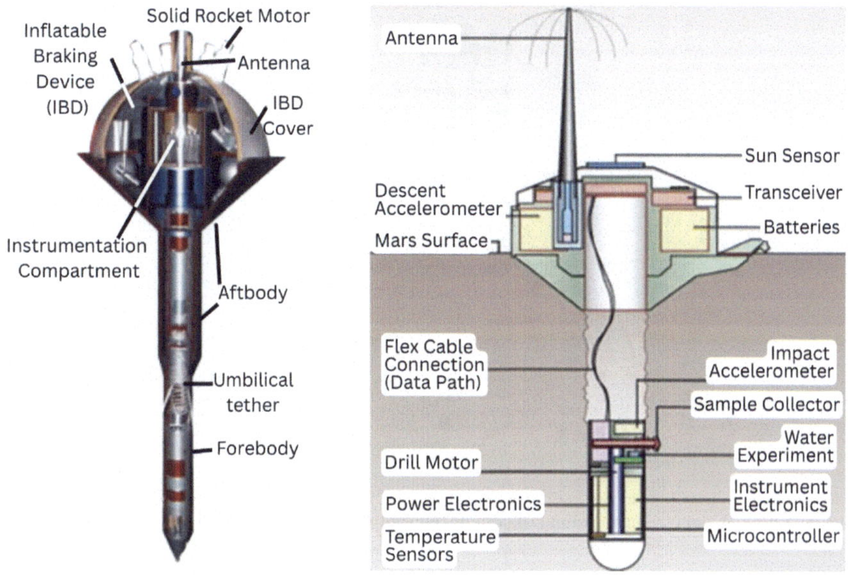

Fig. 2 Cross-section views of Mars 96 (left) and DS-2 (right) penetrators (Ball, 2012)

impact approach represented a different engineering philosophy, trading complexity for reduced stress on the payload. The mission's launch failure prevented its realization, but the concept demonstrated an alternative pathway for penetrator deployment relevant to modern designs (Space Research Institute, 2002). Cross-sectional diagrams comparing the Mars' 96 and DS-2 probes are shown in Fig. 2.

Japan's Lunar-A mission, developed throughout the 1990s and early 2000s, was specifically dedicated to lunar penetrators. It aimed to deploy three penetrators, each around 40 kg (13.5 kg excluding retro-motor), onto the Moon at 285 m/s, targeting depths of 1–3 m. Each carried a seismometer and heat flow probe to investigate the Moon's interior, forming a mini-geophysical network (Mizutani et al., 2005; Nakajima et al., 1996). The operational concept for the Lunar-A mission's geophysical network is depicted in Fig. 3.

JAXA conducted extensive ground testing, firing prototypes into various materials and developing shock absorption techniques, proving instruments could survive impacts up to approximately 10,000 g. However, persistent technical issues led to the mission's cancelation in 2007 before launch. Lunar-A significantly advanced penetrator engineering, particularly for larger probes and sophisticated instruments, but underscored the challenges of mission execution and sustainability.

Interest in lunar penetrator networks was revived in the late 2000s with concepts like the Moon Lightweight Interior and Telecoms Experiment (MoonLITE) mission led by the United Kingdom (Gowen et al., 2008). The MoonLITE concept was later adapted for several mission proposals, including ESA's Lunar Net, Laplace mission to Europa, Tandem mission to Titan and Enceladus, and a Europa Clipper Penetrator

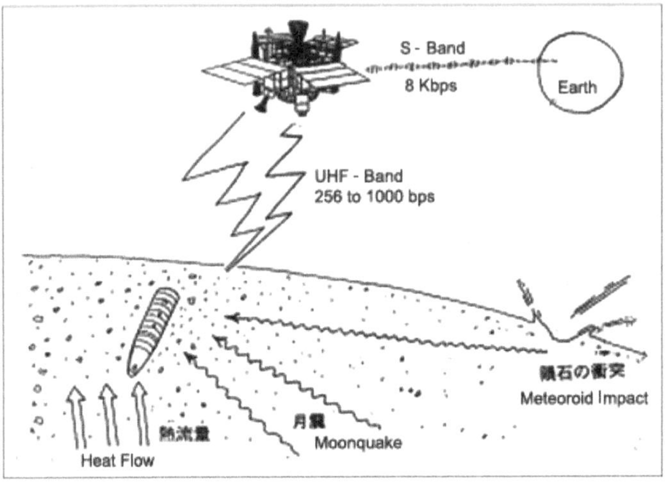

Fig. 3 Lunar-A penetrator mission concept of operation (Mizutani et al., 2005)

mission (Barraclough et al., 2015; Collinson, 2008; Smith et al., 2010). The original MoonLITE mission proposed deploying four penetrators across the Moon, each weighing less than 20 kg, to establish geophysical networks for seismic, heat flow, and geochemical studies at a lower cost than landed stations. MoonLITE also aimed to test communications networking between the probes. The conceptual design for the MoonLITE penetrator and its delivery system is shown in Fig. 4.

These studies refined deployment strategies from orbit and further assessed survivability, including notable tests simulating impacts into sand beds. While neither MoonLITE nor Lunar Net secured funding for flight, they solidified the scientific case for penetrator networks and generated valuable engineering data.

3.3 Current Status and Recent Developments

Beyond these major mission efforts, numerous other concepts have explored penetrator applications. The most comprehensive test suite carried out recently involved tests associated with MoonLITE and its adapted mission concepts. Private initiatives like Explore Mars Inc's ExoLance proposed using small penetrators, along with a crowdfunding campaign, to search for subsurface life on Mars (Cassady, 2014). A sample return penetrator mission concept called Sample Return Systems for Extreme Environments (SaRSEE) was introduced in 2012. Multiple tests conducted until 2015 confirmed the survivability of a hollowed sample-collecting penetrator with a tethered attachment in impacts exceeding 600 m/s (Winglee et al., 2012). Another mission, called Ice Shell Impact Penetrator (IceShIP), aimed to achieve 1–10 m depths, was tested with a comprehensive payload enduring up to 60,000 g impact decelerations

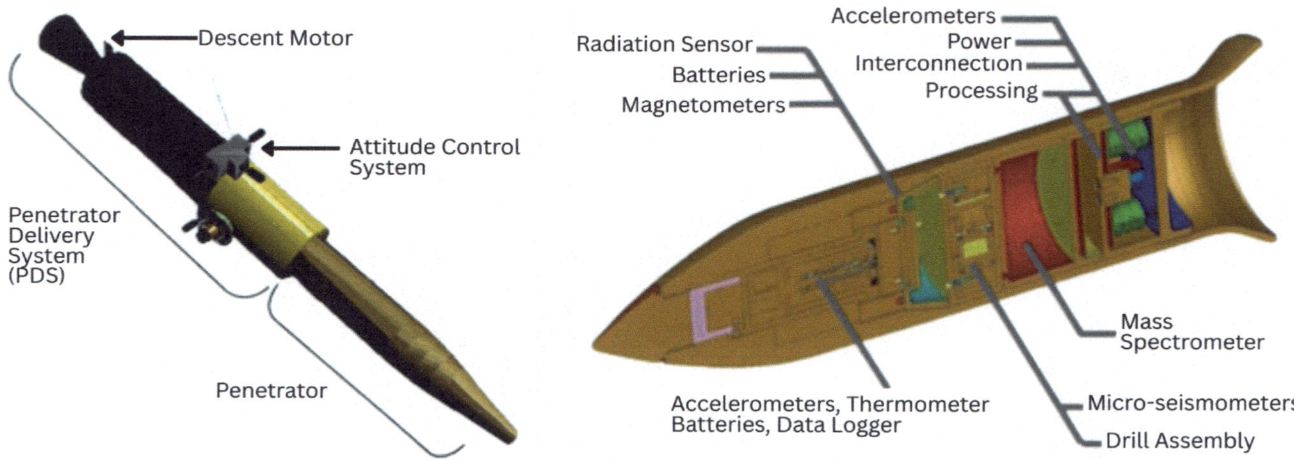

Fig. 4 Concept of MoonLITE penetrator descent module (left) and cross-section view of MoonLITE penetrator (right) (Gowen et al., 2008)

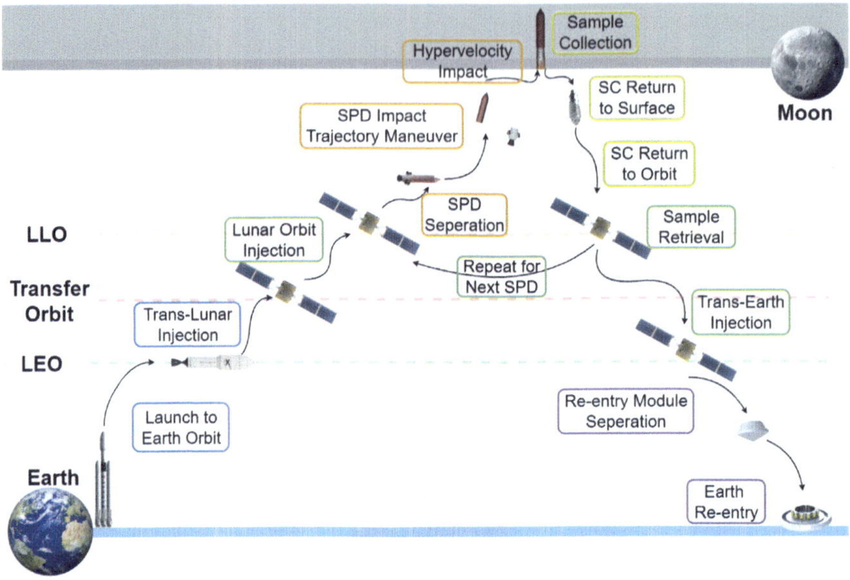

Fig. 5 SPD sample return penetrator mission concept (Landers et al., 2024)

(Raj et al., 2022). More recently, community-driven projects like Nexus Aurora's Soil Penetration Darts (SPD) conceptualized using numerous small darts for lunar sample collection upon impact. Subsequent retrieval could involve a rover or sample return by via direct ascent of its sample collector to a lunar orbit for capture by an orbiter, showcasing innovative hybrid mission architectures and validating their sample collection methods through testing (Landers et al., 2024). This sample return mission concept is illustrated in Fig. 5.

In summary, the history of planetary penetrators reveals high scientific potential hampered by significant engineering hurdles, primarily impact survival and subsurface communication. Failed missions like DS2 and canceled ones like Lunar-A provided critical lessons. However, each attempt has pushed the technological boundaries. Advances in materials, miniaturized electronics, power systems, and communication technology, coupled with innovative concepts like slowed impacts and network deployments, are mitigating past limitations. This legacy of development positions penetrators as increasingly viable tools for future space exploration, including the demanding application of subsurface resource prospecting in space mining. Currently, while no dedicated planetary penetrator missions are in active development by major agencies, research continues.

3.4 Technologies for Modern Penetrators

Overcoming the extreme physical stresses of impact—where deceleration reaches tens of thousands of g's and intense thermal spikes occur—has been the primary driver of technological innovation for penetrators. Following sections highlight several significant advances.

3.4.1 Impact Resistant Materials and Structures

Developments in impact-resistant material have moved beyond traditional high-strength steels and titanium alloys. Ultra-high-strength alloys, advanced composites, nano-enhanced materials, and functionally graded materials offer superior strength-to-weight ratios and toughness. Critically, structural design incorporates sophisticated shock absorption mechanisms. These include internal mechanical metamaterials (such as engineered lattices or honeycombs designed to crush controllably), specialized potting compounds (such as epoxies, foams encasing sensitive components), and nested-shell architectures that dissipate impact energy progressively. SaRSEE tested Hexcel® aluminum honeycomb energy-absorbing structures, showing concertina-style collapse under axial load and dramatic gains in crush strength when filled with epoxy or Kevlar putty (Winglee et al., 2012). Aluminum-foam-filled structures and folding tubes have been extensively tested for penetrator applications (Luo et al., 2021). Advanced manufacturing techniques like 3D printing enable the creation of these optimized, complex structures and have been explored for various space applications (Werkheiser, 2014). Potting and silicon-based shock protection has been widely tested with smart ammunitions and penetrator concepts (Chao et al., 2010; Hopf et al., 2010). Future research continues to explore novel damping materials, potentially including advanced aerogels or optimized metamaterials, alongside improved computational impact modeling to better predict behavior in diverse target materials such as ice, rock, or regolith.

3.4.2 Deployment and Navigation

Controlled deployment and precision navigation is key for a successful penetrator mission. The most common method of penetrator deployment is propelled deployment. Since a penetrator is relatively small compared to a traditional spacecraft, the deployment system is often limited. Various research has been carried out on single-thruster control techniques, which is key for penetrator deployment modules (Raus & Watt, 2012). An equally important aspect is precise navigation. Research ranges from penetrator-specific autonomous navigation and control systems (Badrakalimuthu et al., 2010; Wang et al., 2016) to general-purpose autonomous landing systems like NASA's Autonomous Landing Hazard Avoidance Technology (ALHAT) (Robertson & Carson, 2015).

3.4.3 Thermal Management

Managing the intense heat generated upon impact is essential. Solutions range from passive methods like ablative coatings, insulating materials, heat-spreading materials, and phase-change materials (which absorb heat during melting) to more complex active systems like miniaturized heat pipes or thermoelectric coolers for specific components. Robust thermal design ensures components survive the initial spike and function in the potentially extreme subsurface thermal environment. LunarNet proposed snubber mounts creating a vacuum-insulated bay to cut conductive losses to approximately 0.3 W and radiative losses to 0.5 W, extending penetrator lifetime in cold, conductive regolith (Smith et al., 2010). Other concepts have used thermal sensors in addition to their thermal payloads to measure temperature. Further empirical data on internal temperatures during high-speed impacts into various media are needed to refine thermal designs and validate models, addressing challenges of both impact heating and long-term survival in extreme cold.

3.4.4 Ruggedized Electronics and Miniaturized Instruments

The microelectronics revolution has been pivotal for impact-resistant miniaturized electronics. Micro-Electro-Mechanical Systems (MEMS) technology enables drastic miniaturization of sensors like accelerometers, thermometers, and seismometers, often with inherent robustness against high g-forces. Driven by MEMS and low-power electronics, sophisticated instrument suites can now be integrated into much smaller and lighter penetrators. This includes miniaturized spectrometers, highly sensitive seismometers for interior-structure studies, environmental sensors, and magnetometers. Proposals of MoonLITE, Lunar Net, LunarEx, and IceShIP provide detailed analysis on their miniaturized sensor suites (Raj et al., 2022; Smith et al., 2009, 2010). A modern ~5 kg probe might carry the payload equivalent of a ~15–20 kg probe from decades past, enabling more comprehensive investigations.

3.4.5 Subsurface Communication and Power Systems

Transmitting data from underground remains a significant challenge. Solutions include using low-frequency radio (VLF/LF/HF), which penetrates regolith/ice better than higher frequencies, deploying antennas post-impact, utilizing relay systems via an orbiter or a surface module, or potentially mesh networking between multiple penetrators. DS-2 and MarsNet used UHF links to their orbiters/landers while Lunar Net baselined a UHF/MELACOM system with heritage from Beagle 2 (Zhou et al., 2024). The purpose of the MoonLITE project was to demonstrate high-rate communication links from the lunar surface using its network of penetrators, and therefore had a greater focus on communication bandwidth (Gowen et al., 2008).

Advances in high-energy-density primary batteries coupled with ultra-low-power electronics and efficient power management enable mission durations from weeks to potentially months. For longer missions or high power needs like heating, compact RTGs are becoming more feasible options. Both DS-2 and Lunar-A achieved 275–430 Wh/kg with these cells while Lunar Net assumed $\sim$550 Wh per pack for one-year surface life (Smith et al., 2010).

3.4.6 Sample Collection and Analysis

For resource prospecting with penetrators, the ability to acquire and analyze subsurface samples is key. Technologies include passive impact-driven collection cavities, active mechanisms like small integrated augers or drills, and systems for sample preservation. The SaRSEE design features an internal sample-recovery cavity (SRC) and tether-capture mechanism (Winglee et al., 2012). Yang Gao et al. have presented a micro penetrator concept of about 10 kg in mass that utilizes a Wood wasp reciprocating drill to penetrate deeper into the surface after deployment (Gao et al., 2007). Nexus Aurora's SPD concept features a sample collector device that uses a solid rocket motor and an auger system to ascend to the surface or back to lunar orbit (Landers et al., 2024). Advances in microfluidics enable lab-on-a-chip systems for complex chemical analyses on minute samples, while integrated spectrometers provide detailed compositional data without requiring sample return, offering crucial ground-truth data. All the penetrator concepts that were inspired by the MoonLITE mission concept featured mass spectrometers for in situ volatile characterization (Sheridan et al., 2018).

These developments collectively enhance the operational capabilities and scientific potential of penetrators, enabling longer missions, reliable data return from challenging environments, and sophisticated in-situ analysis or sample acquisition crucial for applications like detailed resource assessment.

4 Penetrator Mission Concepts and Techniques

Building upon the historical development and fundamental capabilities of planetary penetrators, a range of innovative mission concepts and advanced techniques are emerging to significantly enhance their utility, particularly for the demands of space resource exploration and prospecting. Traditional penetrator missions often involved simple, passive deployments. However, future applications necessitate more sophisticated approaches for targeted delivery, coordinated operation, enhanced subsurface interaction, and integration with broader exploration architectures. This section delves into several key concepts and techniques, outlining their operational principles, potential benefits, engineering considerations, and relevance to efficient space mining.

4.1 Traditional Penetrators

Standard penetrators act as passive projectiles, their impact conditions largely dictated by the parent spacecraft's trajectory and gravity. The impact trajectory must be achieved by the cruise or launch spacecraft. While this method offers no maneuverability, it could be ideal for a spacecraft exploring multiple celestial bodies using gravity assists to change its trajectory. The spacecraft might initially enter an impact trajectory with a celestial body, deploy the penetrator, deviate toward a gravity assist trajectory, and cruise toward its next target.

A major advancement to this concept involves equipping penetrators with propulsion capabilities for controlled, precise impacts. The propelled penetrator concept introduces active control, as shown in the deployment sequence in Fig. 6, typically via small rocket motors or braking devices, to adjust velocity and trajectory prior to impact. The primary advantages are precision targeting and impact velocity optimization. Precision targeting allows penetrators to strike specific, high-value locations identified by remote sensing, such as small, shadowed lunar craters potentially harboring water ice, or distinct geological units on an asteroid suspected of containing valuable ores. Impact velocity optimization is crucial for ensuring the probe survives the impact event and achieves the desired penetration depth.

A practical implementation is rocket-assisted delivery, often involving a dedicated carrier or descent module. Instead of free-falling from orbit (which on airless bodies like the Moon results in very high impact velocities in the range of several kilometers per second), the module executes a braking burn using a retrorocket shortly before reaching the surface. This dramatically reduces the impact velocity to a more survivable range, typically around 150–200 m/s. This method was key in past concepts including Mars 96, Lunar-A, and MoonLITE-based penetrators. Reducing impact

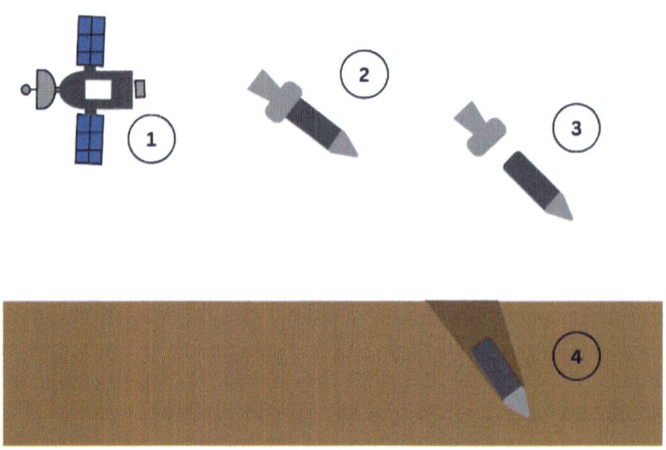

Fig. 6 Propelled penetrator deployment sequence (1. Detachment from Spacecraft, 2. Descent, 3. Propulsion Module separation, 4. Impact)

speeds from several km/s to around 200 m/s significantly lowers peak deceleration loads, greatly increasing survival odds. Furthermore, the controlled deceleration phase allows for trajectory adjustments using small attitude control thrusters on the descent module, refining the impact footprint and reducing the dispersion inherent in purely ballistic deployments, where minor release errors can lead to miss distances of kilometers. This enhanced targeting accuracy significantly improves the return on investment for each deployed penetrator, ensuring probes are more likely to land in scientifically or commercially relevant zones.

Achieving the required targeting precision—potentially on the order of hundreds of meters or less to hit specific small features—remains a significant challenge, especially for deployment from orbit. Research has been conducted on navigation systems for penetrators and single-thruster attitude control techniques. This necessitates continued research into advanced deployment mechanisms, improved carrier vehicle navigation systems, the use of terrain-relative sensing during descent for terminal guidance, and potentially the development of guided penetrators with limited steering capabilities.

The propelled penetrator concept is considered feasible using current technologies, including appropriately sized solid rocket motors and established navigation and guidance algorithms. The main trade-off involves increased mission complexity versus the substantial gains in survival probability, targeting precision, and the ability to deploy effective networks on airless bodies where atmospheric braking is unavailable. For targeted subsurface resource assessment, this trade is often deemed highly favorable.

4.2 Swarm Penetrators

Instead of a single penetrator, a spacecraft can carry a cluster or bundle of smaller penetrators. Similarly, a single descent module can also carry a cluster. After the main deceleration burn, the module can release the individual penetrators, possibly using springs to impart slight dispersion, ensuring they impact distinct but closely spaced locations, forming a localized sensor network in a single deployment event (Fig. 7). While individual penetrators could theoretically carry their own micro-thrusters for final braking, this is often mass-prohibitive for small probes, making the shared braking module for a bundle a more common architecture.

Moving beyond single probes or small clusters, swarm concepts envision deploying penetrators in large numbers to achieve broad spatial coverage, distributed sensing, and high redundancy. Leveraging the inherently lower cost and mass of individual penetrators compared to landers or rovers, a swarm mission could deploy numerous penetrators across a wide region or multiple distinct areas of interest on a planetary body or asteroid. This approach transforms penetrators into a powerful tool for large-scale surveying and characterization. While MoonLITE envisioned a mission with 4 penetrators, a potential concept for a larger swarm was developed by

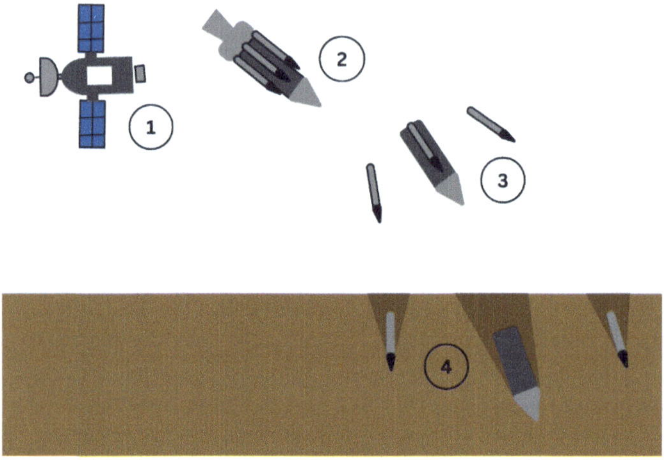

Fig. 7 Swarm penetrators concept with a main penetrator and small penetrator swarm (1. Detachment from Spacecraft, 2. Descent, 3. Swarm separation, 4. Impact)

Space Initiatives Inc, introducing ~1.5 kg "Mote" penetrators, allowing four to fit within a 3-U CubeSat volume (Ahrens et al., 2021).

Swarm deployment offers several advantages. Firstly, a dispersed network of sensors can create high-resolution maps of subsurface properties, such as the distribution and concentration of resources like water ice. While orbital data might identify a promising region, a swarm can pinpoint the richest deposits or variations within that region. Secondly, data collected concurrently from multiple locations enables unique scientific analyses, such as pinpointing moonquake epicenters using arrival time differences in a seismic network (Fig. 8), or understanding the spatial extent and variability of a resource deposit through synchronized measurements. Thirdly, given the inherent risks of high-velocity impact, deploying a large number of probes ensures that mission objectives can still be met even if some units fail. Losing a few probes in a swarm of twenty still yields substantial data return. Fourthly, penetrators within a swarm can be specialized, each carrying different instruments. Together, they provide a comprehensive, multi-faceted dataset covering a wide area, far exceeding the capability of a single penetrator.

Communication coordination is a key challenge, requiring protocols to manage data transmission from numerous sources without interference. A hierarchical network, potentially using one or more penetrators or a dedicated surface module as data relays or hubs, could facilitate communication with an orbiter or Earth. From a cost perspective, swarm missions benefit from economies of scale. Once the penetrator design is finalized and tested, manufacturing additional units is relatively inexpensive compared to the initial development cost.

Fig. 8 Swarm penetrators for triangulated pinpointing of seismic activity

In the context of space mining, swarm penetrators offer an efficient method for comprehensive resource prospecting. Before committing to costly extraction infrastructure, a swarm deployment could rapidly verify orbital reconnaissance data, delineate the extent and grade of deposits, and provide geotechnical information across the target site. These minimize the risks of subsequent mining operations, analogous to terrestrial exploratory drilling campaigns but conducted remotely and rapidly over potentially vast areas.

4.3 Lander-Based Deployment

Penetrators need not solely rely on deployment from orbit or atmospheric entry. They can also be launched from a surface platform, such as a stationary lander or a mobile rover. This concept (Fig. 9) envisions the landed asset acting as a mothership, equipped with a launcher capable of firing penetrator probes into the nearby terrain or to targets some distance away. This approach uniquely combines the stability and resources of a landed platform with the subsurface access capabilities of penetrators.

A primary advantage is the ability to investigate locations otherwise inaccessible to the lander or rover itself, such as the steep walls of a crater, permanently shadowed regions nearby, or hazardous rocky terrain. The lander can remain in a safe, resource-rich location while using penetrators to extend its effective exploration footprint.

Deployed penetrators can be either tethered or untethered. Tethered harpoon like penetrators remain physically connected to the lander via a cable. This allows the lander to supply continuous power and establish a high-bandwidth, reliable data link, enabling longer operational lifetimes and more complex instrumentation within the penetrator (which doesn't need its own substantial power source or long-range transmitter). Furthermore, the tether offers the possibility of retrieving the penetrator after measurement, potentially even extracting a captured sample. However, tethered systems introduce significant complexity, requiring mechanisms to manage the cable during high-speed deployment and potentially during retrieval, making sure

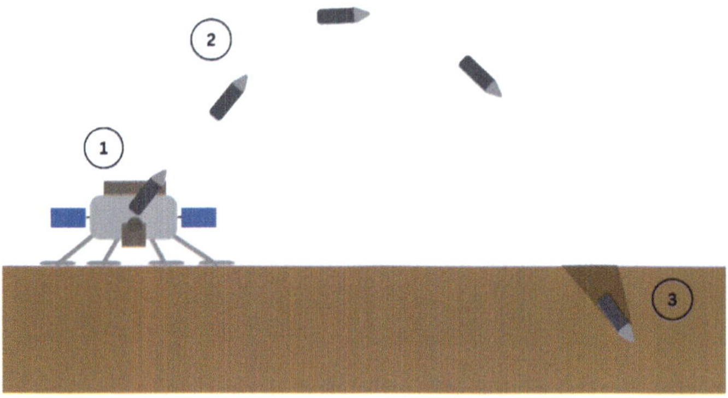

Fig. 9 Landers-based deployment (1. Launch from Lander, 2. Ascent and descent, 3. Impact)

the prevention of snagging or breakage. SaRSEE explored a tethered sample retrieval mechanism (Winglee et al., 2012), while concepts also exist for sample collection via tethered harpoon systems, such as the Comet Rendezvous, Sample Acquisition, Investigation, and Return (CORSAIR) mission (Völk et al., 2018).

Untethered harpoon penetrators are mechanically simpler, essentially projectiles fired toward a target. They operate autonomously after embedding, relying on their own power and wireless communication. This avoids tether-related complexities and allows for greater deployment range, but limits operational duration and data rates, and prevents retrieval. Harpoon launchers themselves have been prototyped, often using pyrotechnic charges, gas cartridges, or spring mechanisms to achieve velocities sufficient for penetrating meters into regolith or ice. Designs must address recoil management to ensure firing does not destabilize the lander, potentially using counter-masses or opposing launches.

Lander-based deployment offers the significant benefit of controlled targeting based on ground-truth assessment. Operators can survey the immediate vicinity using the lander's cameras and instruments before deciding precisely where and when to fire the penetrators, avoiding obstacles and optimizing impact angle for desired depth or target access.

A highly valuable extension is a lander-based harpoon system with sample collection and redeployment. This concept utilizes a tethered harpoon specifically designed to capture a subsurface sample within an internal chamber upon impact. The lander then reels the penetrator back using the tether. The retrieved sample can be transferred to sophisticated analytical instruments aboard the lander. Crucially, the system could reload and redeploy the same harpoon to sample multiple distinct locations around the lander. This capability dramatically enhances scientific return, allowing for comparative analysis of different microenvironments or geological units accessible from a single landing site.

Engineering challenges of such a harpoon system include designing a robust penetrator capable of multiple impacts and retrievals, ensuring sample integrity during the violent collection process, and reliably extracting the probe from potentially cohesive soils. Despite these hurdles, the potential to acquire and analyze multiple subsurface samples from depths inaccessible to typical scoops or drills makes this concept revolutionary for missions focused on astrobiology or detailed geochemical surveys on Mars, icy moons, or asteroids.

4.4 Multi Stage Penetrators

Representing one of the most ambitious extensions of penetrator technology, multi-stage penetrators are conceived as integrated systems capable not only of subsurface investigation but also of returning a physical sample to the surface or even to orbit for collection. This concept goes beyond simple impact-and-measure probes by incorporating distinct functional stages within a single projectile.

A typical architecture involves three stages. An entry/descent stage that is responsible for initial deceleration, potentially using heat shields, parachutes, or retro-rockets, similar to propelled penetrators. This stage is often jettisoned just before impact. An impacting/sampling stage that is the core penetrator body designed to survive impact, penetrate the subsurface, and acquire a sample using mechanisms like internal cavities, scoops, or potentially small drills. An ascent stage which includes a miniaturized propulsion system, likely a compact solid rocket motor, integrated within the penetrator. Once the sample is secured, this stage fires to propel the sample container back out of the impact hole, potentially toward the surface for retrieval by another asset like a rover, or, more ambitiously, launching it into orbit to rendezvous with a waiting spacecraft. A conceptual diagram of a multi-stage sample return to the surface is shown in Fig. 10. Both SaRSEE and SPD mission concepts envisioned such sample return missions (Landers et al., 2024; Winglee et al., 2012).

This concept is particularly appealing for sample return from small bodies like asteroids or Martian moons where escape velocities are low, potentially enabling missions like an asteroid resource prospector to return subsurface samples for detailed terrestrial analysis before committing to larger mining campaigns. For higher-gravity bodies like the Moon or Mars, launching a sample all the way to orbit directly from a penetrator is extremely challenging due to the required delta-V. In such cases, the ascent stage might only lift the sample to the surface or a low altitude for collection by another vehicle.

However, for such a mission, significant engineering challenges must be overcome. Precise synchronization is required to ensure the ascent stage fires only after the penetrator has successfully impacted, collected the sample, and potentially oriented itself correctly. Reliable miniaturization of the ascent rocket motor and propellant to fit within the penetrator's volume and mass constraints is critical. Guidance for the ascent phase, especially for orbital rendezvous, adds further complexity, potentially requiring sophisticated autonomous navigation or pre-positioning of the collection spacecraft.

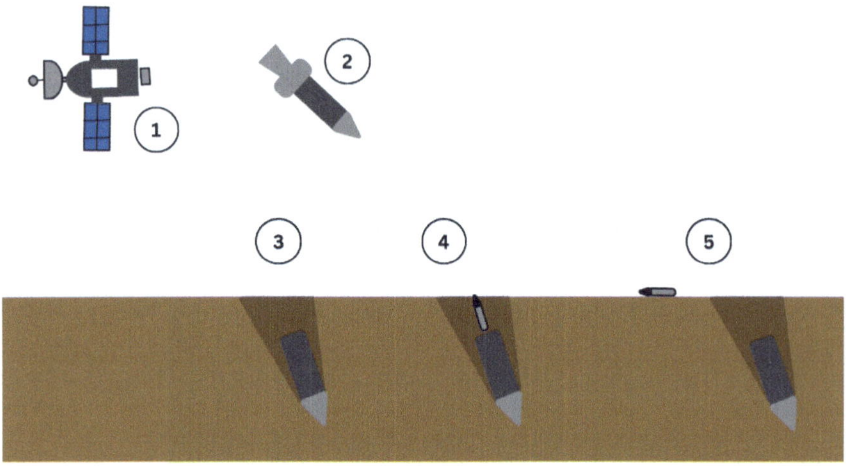

Fig. 10 Multi-stage sample return mission (1. Detachment from spacecraft, 2. Descent, 3. Impact, 4. Sample collector ascent to the surface, 5. Sample collector reaching the surface)

Another concept is to use a forebody that impacts the surface at a higher velocity to create a deep impact hole for the penetrator to impact the same location at a significantly lower velocity (Fig. 11). This approach reduces the challenges of safeguarding instruments from a hyper-velocity impact.

For space mining, multi-stage sample return penetrators could enable rapid, relatively low-cost preliminary assessments of planetary bodies. A prospecting mission could deploy such a penetrator to a target asteroid, retrieve a subsurface sample,

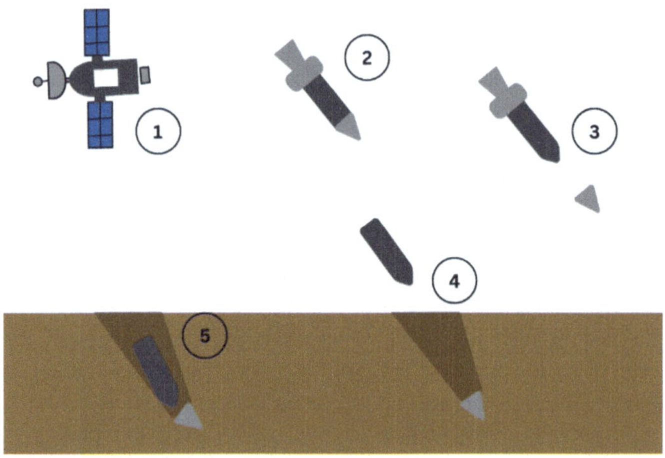

Fig. 11 Penetrator with forebody impactor concept (1. Detachment from spacecraft, 2. Descent, 3. Release of forebody and deceleration of main penetrator, 4. Forebody hyper-velocity impact, 5. Main penetrator low velocity impact)

and return it to Earth for detailed mineralogical and compositional analysis before committing to a larger mining campaign. While still largely conceptual, with prototypes like the Nexus Aurora's SPD exploring staged propulsion, multi-stage penetrators offer a potential pathway to highly capable, end-to-end resource reconnaissance systems.

4.5 Surface Module Deployments for Power and Communication Support

Individual penetrators, especially those deployed in swarms or intended for long-duration monitoring, often face limitations due to their small size, constraining their power reserves and communication capabilities. A complementary approach involves deploying dedicated surface modules in conjunction with penetrators to act as local infrastructure hubs (Fig. 12). These modules remain on the surface, providing essential support services.

A typical surface support module might be equipped with larger solar panels or a radioisotope power source, a high-gain antenna for communication with Earth or orbiters, precise timing references, and potentially navigation aids.

A surface module may be connected to penetrators via a tether, as envisioned in DS-2 and Mote penetrator concepts (Ahrens et al., 2021; Smrekar et al., 1999). Even for untethered probes, if designed for retrieval or proximity operations, the module could potentially offer additional capabilities. Deployed centrally within an area where penetrators are impacting, the station could significantly enhance the network's capabilities.

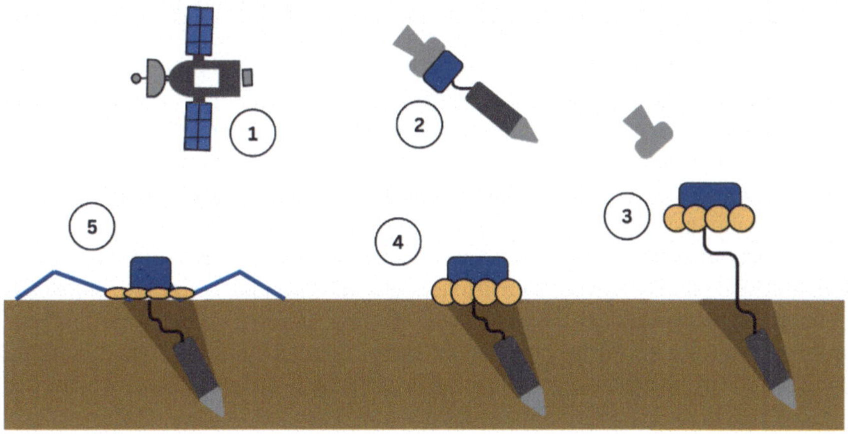

Fig. 12 Surface module and penetrator concept (1. Detachment from spacecraft, 2. Penetrator separation and surface module deceleration, 3. Penetrator impact and surface module airbag deployment, 4. Surface module landing, 5. Surface module mechanism deployment)

The primary advantage of a surface module is the ability for it to provide continuous power to the penetrator. Using solar panels, radioisotope generators, or larger batteries, these modules can power one or more penetrators, while the penetrators themselves might only need a smaller battery for critical functionalities. Similarly, penetrators could use low-power, short-range radios to transmit data to the nearby surface module. The module, with its more powerful transmitter and larger antenna, could then aggregate this data and relay it efficiently over long distances. The surface module could also serve as a central data logger and potentially perform preliminary data processing. Furthermore, it could help penetrators determine their position or synchronize measurements across the network by emitting reference signals. By carrying its own instruments, the module could provide contextual data about surface conditions relevant to interpreting penetrator measurements. Merging the strengths of landers with penetrators, this hybrid approach acts as a significant force multiplier for resource exploration missions.

4.6 Advanced Sampling Mechanisms and Post-Impact Techniques

Beyond the major mission concepts, numerous specific technological innovations focus on enhancing how penetrators acquire samples, interact with the subsurface after impact, and communicate their findings. These advanced mechanisms and techniques refine penetrator performance for resource exploration and scientific investigation.

4.6.1 Internal Sampling Mechanisms and Sample Collector Devices

Improving upon simple cavity collection, advanced designs incorporate features like one-way valves or spring-loaded trap doors at the sample inlet to prevent loss after capture. Internal geometries are optimized to encourage material flow while minimizing excessive compaction that could hinder analysis. An example of this is SPD project's aftbody pyro collapsible door system that allows the soil debris in the impact hole to fall into a cavity which contains its sample collector (Fig. 13).

A further advancement to the sample collector is a device with return capabilities. Both SaRSEE and SPD projects envisioned returnable sample collectors. SaRSEE's device (Fig. 14) was attached to a tether, collecting samples via cutaways and feed ports (in the nose cone) during penetration. The SPD collector was designed to ascend to the surface using a drill auger and a motor, collecting samples along the way, and then use a solid rocket motor for orbital ascent and capture.

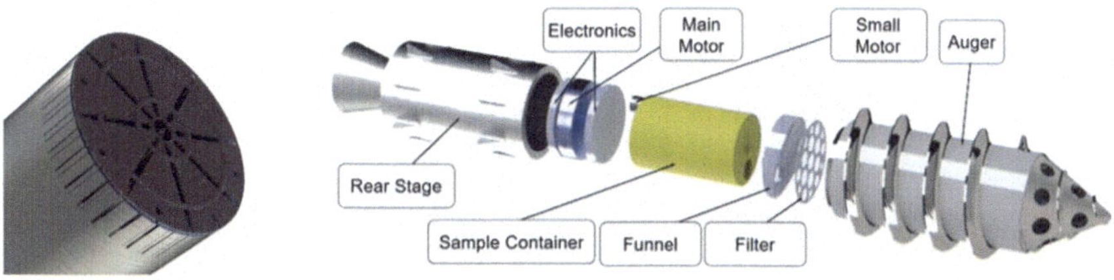

Fig. 13 SPD rear end (left) and SPD sample collector with ascent capabilities (right) (Landers et al., 2024)

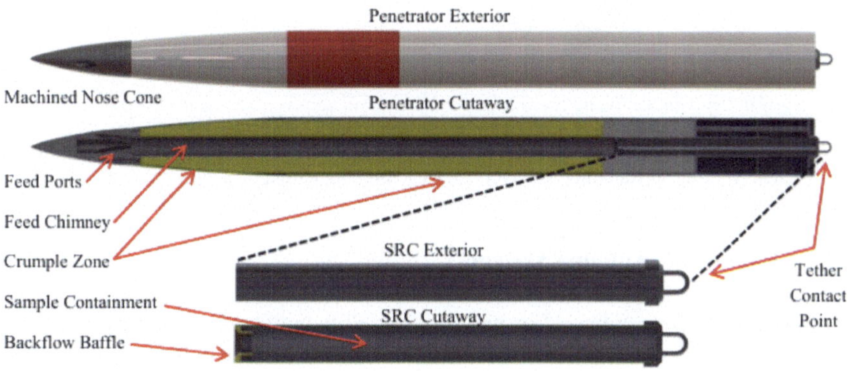

Fig. 14 SaRSEE penetrator (Winglee et al., 2012)

4.6.2 Melting Probes

Particularly relevant for icy bodies (like Europa, Enceladus, Mars poles), this technique supplements or replaces kinetic energy with thermal energy. A penetrator might incorporate a heated tip (powered electrically or by radioisotopes) that activates post-impact. This concept is illustrated in Fig. 15. This allows the probe to melt its way deeper into the ice under gravity over time. Such thermal impactors could potentially reach depths of meters or tens of meters, accessing regions below the heavily processed surface layer, crucial for astrobiology or climate studies. Several research efforts have explored mission concepts using similar thermal probe methods for Europa and other icy bodies (Biele et al., 2010). Hybrid designs might combine impact penetration with subsequent thermal melting or incorporate small mechanical drills to handle non-ice inclusions. Challenges include thermal efficiency, ensuring descent in low gravity, and potential interactions between meltwater and vacuum environments.

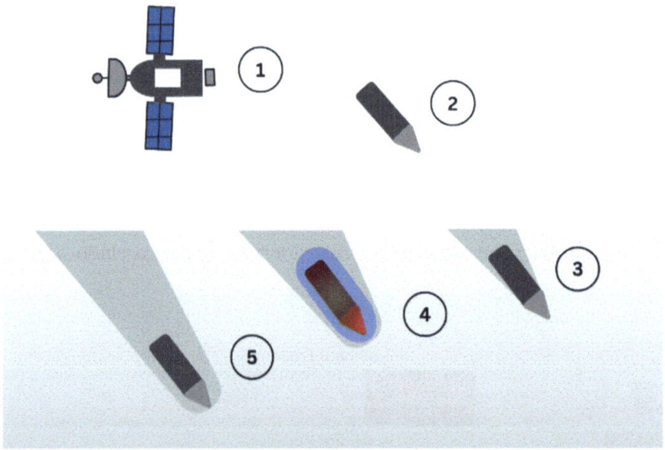

Fig. 15 Thermal drilling and melting probe concept (1. Detachment from spacecraft, 2. Descent, 3. Impact, 4. Melting surrounding ice layers, 5. Reaching higher depths)

4.6.3 Deployable Antenna Systems

Communicating from a buried or partially buried penetrator with limited power is challenging. Innovations adapted from small satellites include compact, deployable antennas such as spring-loaded or helical antennas. These larger antennas improve the link budget, allowing for more reliable communication over longer distances or at higher data rates using lower transmission power compared to fixed, body-mounted antennas (Moulder, 2021).

Alternative concepts include firing a small harpoon-like dart carrying a wire antenna out of the ground from the embedded penetrator to establish a clearer transmission path (Fig. 16). Robust, shock-tolerant deployment mechanisms are key to the feasibility of such deployable instruments.

These advanced sampling, interaction, and enabling technologies transform penetrators from simple impactors into sophisticated subsurface probes. Combined with the novel mission architectures, they significantly enhance the capabilities available for detecting, characterizing, and ultimately enabling the sustainable utilization of space resources.

4.7 Future Outlook, Sustainability, and Challenges

The development and deployment of these novel penetrator concepts are central to the future of efficient space resource prospecting and utilization. Their successful implementation, however, hinges on addressing key remaining challenges and embracing forward-looking considerations.

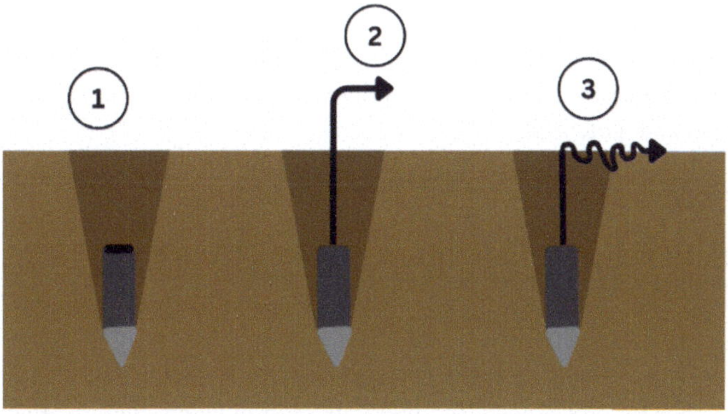

Fig. 16 Harpoon antenna deployment (1. Impact, 2. Antenna release, 3. Antenna deployment)

Beyond the ongoing work on survivability and communications, significant challenges include financial and logistical hurdles. Penetrator missions often require economies of scale, but limited flight heritage can inflate development costs and hinder funding acquisition compared to established lander/rover paradigms. Successful, cost-effective technology demonstrations are crucial. Furthermore, mitigating risks from environmental unknowns, such as unexpected variations in surface hardness or subsurface layering, requires robust design margins or the development of adaptive systems that can adjust performance based on real-time conditions.

As space activities expand, sustainability is paramount. Penetrators inherently offer advantages due to their lower mass and minimal surface disturbance compared to larger systems. Future designs should enhance this by exploring recoverable components, self-remediating features, and adaptive sampling to maximize scientific/resource return while minimizing impacts and energy use. Integration with renewable power sources for longer-duration surface modules is also key.

The deployment of resource-prospecting penetrators will inevitably intersect with the evolving landscape of space policy and governance. Demonstrating responsible practices in early missions will be vital. International cooperation and open data sharing regarding resource findings could foster transparency and prevent conflicts, ensuring space resources offer broad benefits. This necessitates continued dialogue on international coordination, adaptive regulatory frameworks, and ethical considerations.

Penetrators are most effective when integrated into broader exploration architectures, serving as precursors to de-risk landing sites, acting as nodes in distributed networks, or complementing orbital and surface assets. Realizing their full potential requires strong interdisciplinary collaboration between scientists and engineers throughout mission design, deployment, and data interpretation. Analogue field testing remains crucial for refining operational concepts, especially for complex tasks like tethered retrieval.

Despite the challenges, the unique advantages of planetary penetrators—providing rapid, direct subsurface access across wide areas with relatively lower mission complexity and cost—position them as indispensable tools for the next era of sustainable space exploration and resource utilization. Continued focused research, technological maturation, and successful flight demonstrations are key to unlocking this transformative potential.

5 Simulation-Based Validation of Novel Concepts Using DEM

This section explores the feasibility and performance characteristics of several novel penetrator concepts proposed in the previous section using numerical simulations conducted via the Discrete Element Method (DEM). DEM has been successfully applied to many impact penetration problems as discussed by Yu (2009), and is particularly well-suited for modeling the complex, discontinuous nature of granular material interactions during high-velocity impacts.

5.1 Methodology and Simulation Setup

The foundational physics and parameters for the DEM model were adapted from the comprehensive study by Chung et al. (2017), prepared under contract to the Department of Defense Strategic Environmental Research and Development Program. The study focused on physics-based prediction of projectile penetration in granular media and validated numerical models against geotechnical centrifuge experiments. While Chung et al. utilized LS-DYNA software package, the underlying physical principles are transferable. Following their methodology, a Hertz-Mindlin (no slip) contact model with a Standard Rolling Friction model was employed for particle interactions in the alternative software package Altair EDEM (Altair, 2024).

Parameters defining the granular media (regolith simulant) and the penetrator were derived from the 26.5-g scaled centrifuge test scenario (Test Scheme 1) described by Chung et al. This involved using a scaled numerical representation of aluminosilicate spheres (0.93 $\pm$ 0.08 mm non-scaled diameter) as the granular medium. Key parameters adapted include material properties of steel (including an adapted Young's modulus for steel of 200 GPa), particle size distribution (fixed size spherical), particle density (3430 kg/m^3), elastic properties such as Young's modulus (550 MPa), and Poisson's ratio (0.27), and friction coefficients for particle-to-particle and particle-to-penetrator interactions.

Recognizing the differences between EDEM and LS-DYNA implementations and aiming to streamline simulations for this paper's specific goals, some minor simplifications and adaptations were made to the Chung et al. model setup. The coefficient of restitution between particles was derived as 0.046 based on the normal damping coefficient calculation methodology presented. To approximate the string-pot damping effect used by Chung et al. for boundary energy absorption, a simplified approach was implemented using container wall materials with a very low coefficient of restitution (0.001) and standard friction parameters, omitting the explicit circumferential geometric stress.

While Chung et al. noted their LS-DYNA model predicted penetration depths approximately 17% deeper than corresponding centrifuge tests for Test Scheme 1, preliminary simulations conducted using the adapted EDEM setup indicated results that closely align with the centrifuge experimental data (based on the appropriate scaling factors). Given that the primary objective here is not absolute predictive accuracy but reasonable performance estimations and, more importantly, a comparative analysis between the proposed novel concepts, this level of agreement was considered sufficient. The assumptions made for this paper's implementation did not include factors that would significantly impact the results' accuracy (Brandner & Koch, 2023). The adapted model provides a consistent baseline for evaluating the relative performance differences introduced by variations in penetrator design and deployment strategy.

5.2 *Simulation Series and Data Acquisition*

A systematic simulation series was executed for concept validation. Two Soil Bed configurations and two penetrator designs were used. The validated regolith simulant with 22.525 mm diameter DSE was used throughout the simulations.

- **Baseline Parameter Sweep**
  - *Varying Penetrator Mass:* 100, 150, 200, and 254 kg
  - *Varying Impact Velocity:* 70, 140, 210, and 280 m/s
  - *Penetrator:* Diameter (238 mm), Length (834 mm)
  - *Soil Bed:* Diameter (1.16 m), Height (6.89 m)
- **Lander-Based Deployment Simulation**
  - *Varying Angle:* 30, 40, and 50°
  - *Varying Impact Velocity*: 20, 30 m/s
  - *Penetrator:* Diameter (90 mm), Length (446 mm), Mass (7.5 kg)
  - *Soil Bed:* Bottom Diameter (1.7 m), Top Diameter (2.3 m), Height (3 m)
- **Forebody Penetrator Simulation:** Representing a potential forebody component of a multi-stage system with a penetrator shell and a small penetrator.
  - *Shell:* Diameter (238 mm), Length (834 mm), Mass (35 kg)
  - *Penetrator:* Diameter (90 mm), Length (446 mm), Mass (10 kg)
  - *Soil Bed:* Bottom Diameter (1.7 m), Top Diameter (2.3 m), Height (3 m)
- **Crater Formation Analysis:** Two simulations (Baseline with 70 m/s and shell with 350 m/s) were run for extended durations to allow for the complete formation

and stabilization of the impact crater, providing data on ejecta dynamics and final surface morphology.

Comprehensive data was collected for each simulation, including the full deceleration-time history of the penetrator, final penetration depth, soil displacement field and crater characteristics, deceleration forces, and kinetic energy dissipation among soil layers. Visuals from these simulation setups are shown in Fig. 17.

The extensive parametric study presented by Chung et al. (2017), which systematically varied interparticle friction (0.4–0.54), particle-projectile friction (0.57–0.65), and impact velocity (70–110 m/s) across different test schemes, provides valuable context. Their results quantify the sensitivity of penetration depth to these parameters, offering supplementary data that complements the expected trends from the simulations conducted here and aids in interpreting the underlying physics of the impact.

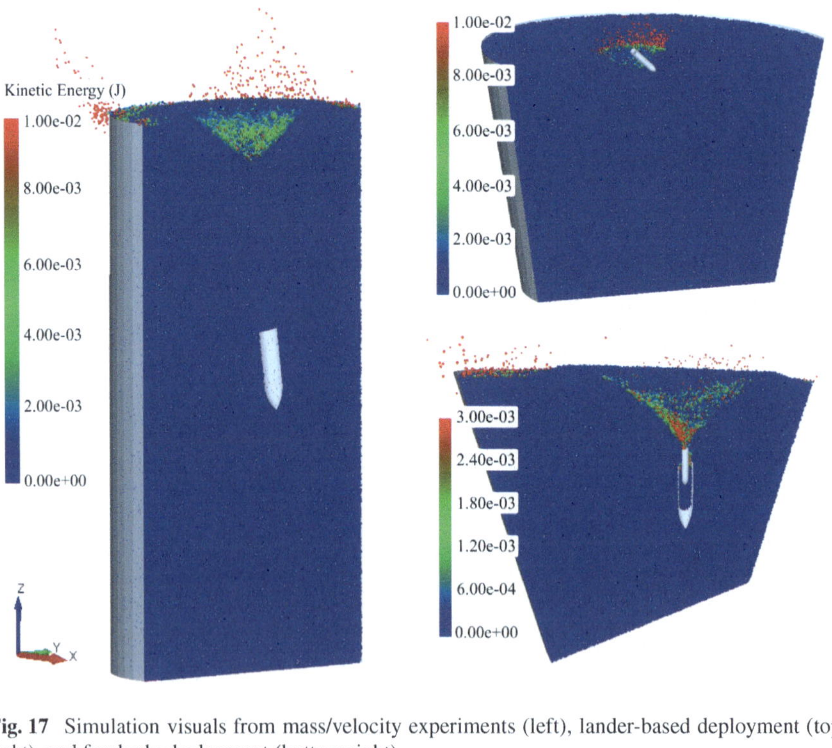

Fig. 17 Simulation visuals from mass/velocity experiments (left), lander-based deployment (top right), and forebody deployment (bottom right)

5.3 Results Interpretation and Concept Viability

The simulation results provide quantitative insights into the performance characteristics and viability of the novel penetrator concepts.

5.3.1 Impact Velocity Effects

Analysis of impact velocity variations (Fig. 18) reveals a strong positive correlation between initial velocity and maximum penetration depth. Increasing impact velocity from 70 to 280 m/s increased penetration depth from 3.2 to 7.0 m. This relationship shows a nearly linear trend, indicating that penetration capability scales predictably with impact energy. Velocity profiles show higher initial velocities generate proportionally higher deceleration forces (exceeding 3.5×10^6 N for 280 m/s). All penetrators decelerate to near-zero velocity within approximately 0.125 s, though higher-velocity impacts require longer distances. Force versus depth profiles demonstrate that higher-velocity impacts maintain significant forces throughout their longer penetration paths, whereas lower-velocity impacts experience more rapid force diminishment with depth. This sustained force at depth for high-velocity penetrators suggests they can effectively access deeper regolith layers while retaining sufficient energy for potential sampling or instrument deployment operations at their terminal positions.

5.3.2 Mass Effects

Simulations show mass significantly influences depth (Fig. 19). Increasing mass from 100 to 254 kg proportionally increased penetration from 1.70 to 3.20 m, also following a linear trend. Unlike the variable velocity cases, all mass variants exhibit similar initial velocity profiles, but heavier penetrators maintain their momentum deeper into the regolith, resulting in extended penetration distances. Force profiles reveal that all mass variants experience comparable peak forces during initial impact, approximately 400,000 N, but heavier penetrators sustain significant forces over longer durations and greater depths. The force versus depth data indicates that while initial force peaks are similar across mass variants, heavier penetrators maintain substantial forces at depths where lighter variants have already come to rest, suggesting greater potential for deeper subsurface operations.

5.3.3 Lander-Based Deployment Analysis

The lander-based deployment simulations (Fig. 20) evaluate penetrator performance at lower velocities (20–30 m/s) and varying impact angles (ranging from 30 to 50° from vertical). The results show that at 20 m/s, increasing the impact angle from 30 to 50° reduces the maximum penetration depth from 0.32 to 0.23 m. Similarly, at 30 m/s,

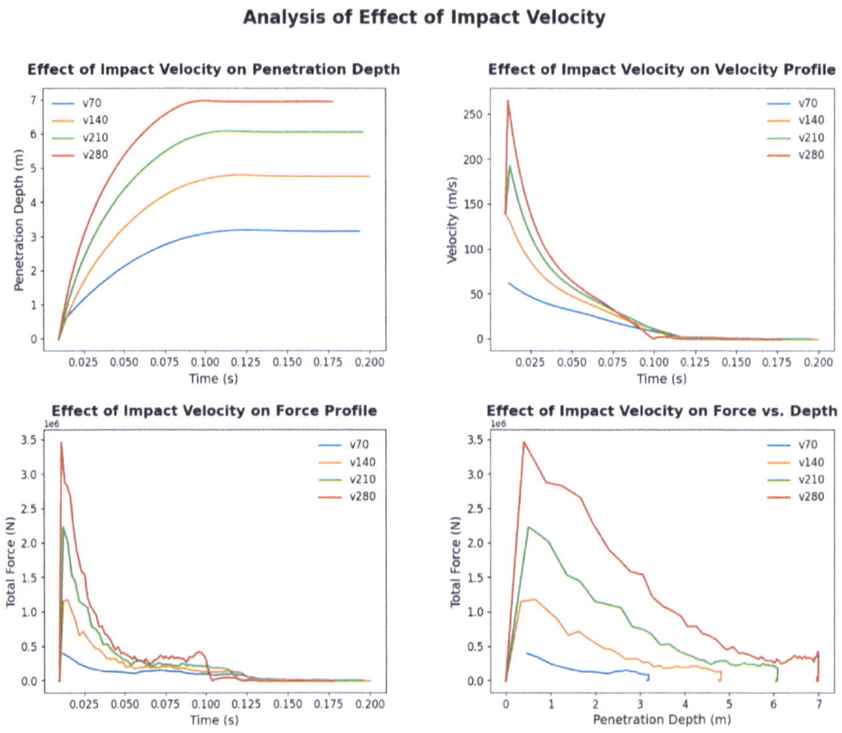

Fig. 18 Effect of impact velocity

the same angular increase results in penetration depth reduction from 0.36 to 0.25 m. These results indicate that higher deployment velocity can partially compensate for non-optimal impact angles, but all lander-based deployments achieve substantially shallower penetration compared to the high-velocity scenarios due to the smaller-sized penetrator (90 mm) and lower mass (7.5 kg). The force profiles for these shallow penetrations show significantly lower peak forces compared to high-velocity impacts, indicating reduced structural requirements for lander-deployed penetrators but also limited subsurface access capabilities.

5.3.4 Forebody Penetrator Performance

This simulation (Fig. 21) examined a smaller, potentially detachable component. The penetrator achieved a maximum penetration depth of 1.52 m with a peak force of 27,649 N. The initial velocity rapidly decreases to near zero within approximately 0.075 s, with the force profile showing multiple distinct peaks indicating complex penetration dynamics due to the forebody being present in a deeper soil layer. These

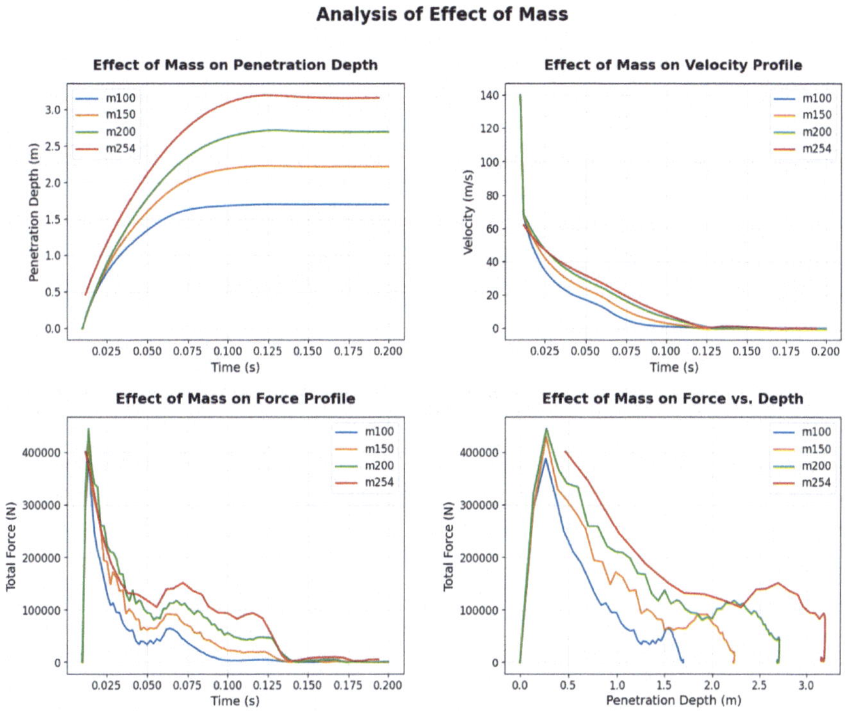

Fig. 19 Effect of mass

results suggest that a forebody penetrator configuration can achieve meaningful penetration depths despite its smaller mass, providing a viable strategy for multi-stage deployment scenarios where initial anchoring or preliminary sampling is desired before deeper penetration by a main penetrator body.

5.3.5 Crater Formation and Ejecta Dynamics

The crater formation analysis reveals significant differences between standard penetrator impacts and multi-stage deployment approaches. Standard penetrators create deeper, more symmetric craters with concentrated ejecta patterns, while multi-stage configurations produce more complex crater morphologies with wider ejecta distribution (Fig. 22).

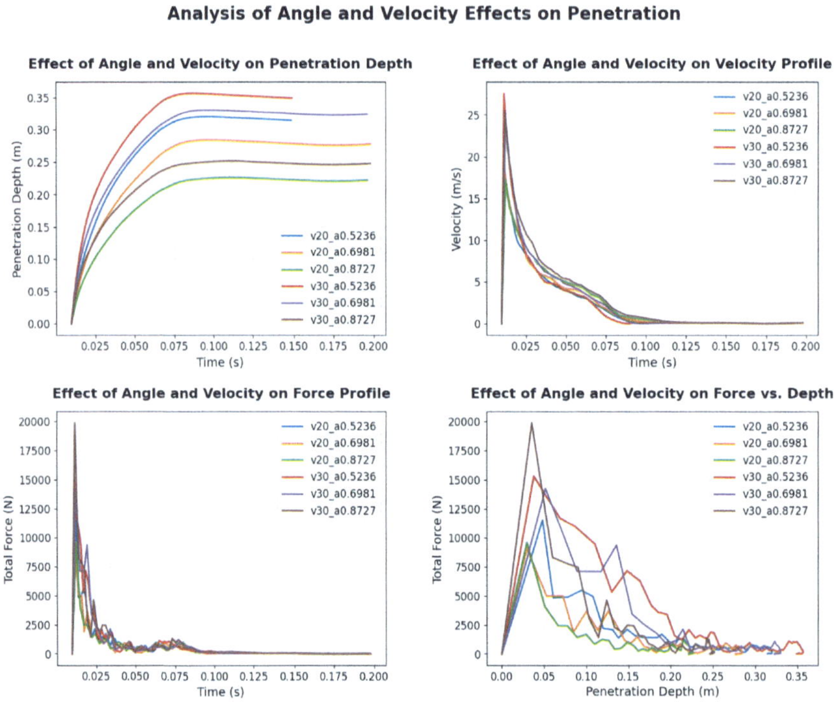

Fig. 20 Landers-based deployment analysis

5.4 Penetrator Technology Viability for Space Mining

The comprehensive simulation results strongly support the viability of the proposed penetrator concepts for space resource identification and prospecting. The data confirm that high-velocity penetrators can reach depths of upto 6–7 m, sufficient to access subsurface ice deposits and mineral layers typically inaccessible via remote sensing. The lander-based deployment simulations validate the feasibility of lower-velocity penetrator deployment, enabling integration with existing landing systems where propulsive capability for high-velocity impacts may be limited.

The near linear relationships between penetrator mass and achievable depths, and between impact velocity and penetration depth, provide quantitative guidance for mission design tradeoffs. For resource-constrained missions, increased impact velocity can effectively compensate for reduced penetrator mass. The forebody penetrator results demonstrate that specialized geometries and multi-stage designs can achieve effective penetration despite lower mass, enabling more sophisticated sampling strategies than monolithic penetrators.

The consistent relationships observed across parameter variations suggest that penetrator performance in granular media is highly predictable, reducing mission

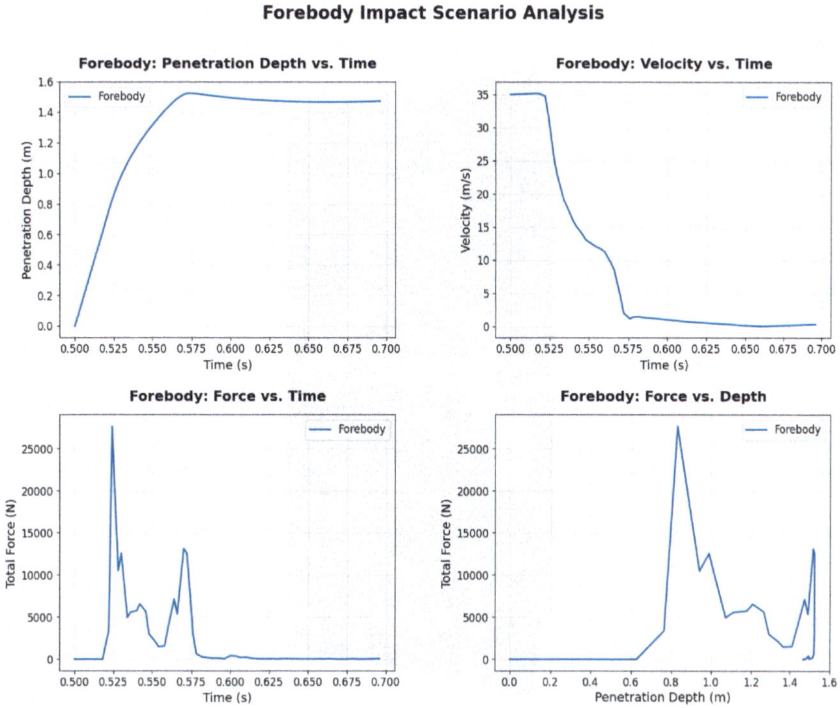

Fig. 21 Forebody deployment

risks associated with deployment in unknown subsurface environments. The demonstrated ability to tune penetration depth through controlled variations in mass and velocity supports the concept of deploying multiple penetrators at different depths, creating a three-dimensional subsurface sensing network for comprehensive resource characterization.

5.5 Limitations

Despite the promising results, several limitations of the current simulation study must be acknowledged. The model utilizes idealized spherical particles, whereas natural regolith contains irregular shapes affecting packing density and friction behavior. The parameter adaptations between modeling environments introduce approximations that may affect absolute accuracy, though the relative performance differences between configurations remain valid.

Additional limitations include simplified treatment of regolith heterogeneity, limited exploration of penetrator geometry variations beyond standard shapes, and focus primarily on impact and penetration phases without detailed modeling of

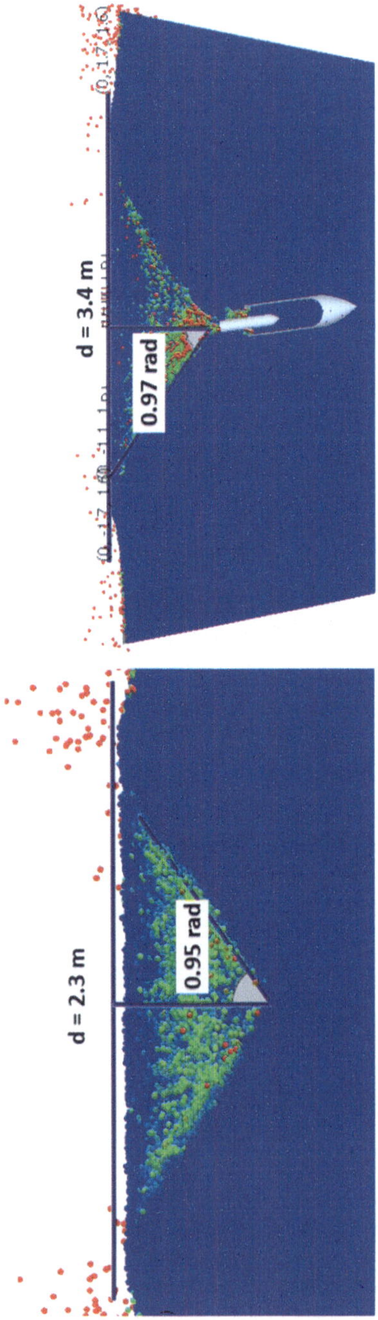

Fig. 22 Crater creation: normal penetrator impact at 140 m/s (left), forebody (350 m/s) and penetrator (35 m/s) impact (right)

post-penetration operations. The simplified treatment of deployment dynamics, particularly for the multi-stage configurations, also represents an area for future refinement.

Nevertheless, the simulations provide a consistent baseline for evaluating relative performance differences between various penetrator designs and deployment strategies, offering valuable insights for future mission planning and technology development in the emerging field of space resource utilization.

6 Legal, Ethical, and Sustainability Considerations

Advancing planetary penetrator technology for space resource prospecting requires careful consideration of legal frameworks, ethical implications, and sustainability practices. As tools for subsurface access and potential resource characterization, penetrators interact directly with celestial bodies, triggering obligations under international space law and raising questions about responsible exploration and utilization. Ensuring penetrator missions are conducted sustainably and ethically is crucial for the long-term viability of space resource activities.

6.1 Legal Framework for Space Resource Utilization

The cornerstone of international space law is the 1967 Outer Space Treaty. It establishes that "outer space, including the Moon and other celestial bodies, is not subject to national appropriation by claim of sovereignty" and affirms freedom of exploration and use for peaceful purposes (United Nations, 1967). While the Treaty does not expressly forbid resource extraction, its Article IX requires all activities to be conducted with "due regard" for other nations' interests and to avoid "harmful contamination" of celestial bodies or adverse changes to the Earth's environment.

In practice, several nations have enacted national laws recognizing rights in extracted space resources without asserting territorial sovereignty. The U.S. Commercial Space Launch Competitiveness Act grants U.S. citizens rights to any asteroid or space resource they recover, while explicitly disclaiming sovereignty over celestial bodies (U.S. Congress, 2015). Luxembourg's 2017 Space Resources Law likewise provides that "space resources are capable of being owned" and creates an authorization regime for their commercial extraction (Grand Duchy of Luxembourg, 2017). Japan's 2021 Act confers ownership of mined resources to licensed operators by intent and possession (Government of Japan, 2021), and the UAE's 2023 Regulatory Framework entitles operators to exercise ownership rights under national law over resources they extract (UAE Space Agency, 2023). The non-binding Artemis Accords further affirm that extraction "does not inherently constitute national appropriation under Article II of the Outer Space Treaty" (NASA, 2020). By contrast, the 1979 Moon Agreement declares that neither the surface nor subsurface of the Moon, or natural resources in place, "shall become property of any State…or of any natural person" (United Nations, 1979).

Under the 1972 Liability Convention, launching States are absolutely liable for damage caused by their space objects (United Nations, 1972), and the 1975 Registration Convention mandates transparent registration of all launched objects to ensure accountability (United Nations, 1975). Together, these treaties and national laws permit penetrator missions for exploration and extraction, provided they honor non-appropriation, due regard, non-interference, transparency, and international responsibility.

6.2 *Sustainability of Penetrator Based Operations*

A truly sustainable penetrator mission begins with planetary protection—the internationally agreed practice of preventing both forward contamination of target bodies and backward contamination of Earth. Under the Committee on Space Research (COSPAR) policy, direct-contact missions, including landers and penetrators to bodies of biological interest such as Mars, Europa, or Enceladus, are classified Category IV. Category IVa vehicles not carrying life-detection payloads must limit total surface bioburden to $\leq 3 \times 10^5$ spores, and an average of ≤ 300 spores m^{-2}. When probes search explicitly for extant life or access special regions such as subsurface zones with transient liquid water, more rigorous heat or chemical sterilization is required to achieve at least a 10^4 fold reduction in viable organisms. Sample-return penetrators fall under Category V, mandating fully contained return hardware and multi-stage bioshielding, with pre- and post-launch verification of sterility (Kminek et al., 2017).

Beyond biology, environmental sustainability demands minimization of mission-generated debris. Agencies follow NASA's Orbital Debris Mitigation Standard Practices (U.S. Government, 2019) and ISO 24113:2023 "Space Debris Mitigation Requirements" to limit on-orbit lifetime of expended stages and ensure post-mission disposal (ISO, 2023). On the surface, hardware footprints are minimized through miniaturization and precision targeting, and by selecting non-toxic or rapidly degradable materials to avoid long-term chemical leaching.

Finally, open data sharing, modeled on the deep-sea regime of the UN Convention on the Law of the Sea, ensures non-monetary benefits like high-value scientific data are disseminated widely. The International Seabed Authority's framework for sharing prospecting results provides a valuable analogy (United Nations, 1994).

Ultimately, responsible penetrator missions require adherence to international law, rigorous implementation of planetary protection, minimization of environmental impacts through thoughtful design and operation, and a commitment to transparency, international cooperation, and ethical principles. By embracing these practices, penetrator technology can serve as a valuable tool for sustainable space resource discovery and utilization, contributing positively to humanity's future in space.

7 Conclusion

In conclusion, planetary penetrators are a pivotal technology for space resource exploration, bridging the gap between orbital sensing and costly landed missions by offering direct, cost-effective subsurface access. This capability is crucial for identifying and verifying valuable deposits like water ice, complementing traditional methods by providing essential ground-truth data and scouting locations for future missions. Significant technological advancements in materials, miniaturization, power, and communications have boosted their feasibility and scientific return, enabling risk-distributed resource verification before large-scale investments.

Key challenges remain, including improving impact survival rates, enhancing power and subsurface communication, and further miniaturizing analytical instruments. Sustainable deployment, addressing environmental impacts, and adhering to regulatory frameworks are also essential.

Looking ahead, penetrators are poised to become integral to space resource activities, initially for verifying resources like lunar ice for programs like Artemis, and potentially evolving into networked systems across the solar system for comprehensive resource assessment. As key enablers for sustainable space expansion, penetrators, developed and deployed responsibly, offer immense potential to unlock vital extraterrestrial resources, representing small impacts that could lead to giant leaps in humanity's ability to operate beyond Earth.

References

Adams, G., Amundsen, R., Beardsley, R., Cash, R. C., Knight, T., & Martin, J. (1986). *Study of comet nucleus gamma-ray spectrometer penetration system*. NASA.

Ahrens, C. J., Paige, D., Eubanks, T. M., W. B., Mesick, K. E., Zimmerman, W., Price, S., et al. (2021). Small penetrator instrument concept for the advancement of lunar surface science. *The Planetary Science Journal*.

Altair. (2024). *Altair EDEM 2024*. Retrieved from Welcome to Altair EDEM https://2024.help.altair.com/2024/edem/index.htm

Aubert, Z. (2025, February 04). *AstroForge reveals asteroid mining mission target & future launch partner*. Retrieved from The Launch Pad Network https://tlpnetwork.com/news/america/astroforge-reveals-asteroid-mining-mission-target-and-launch-partner

Badrakalimuthu, A. S., Lappas, V. J., & Ha, J. C. (2010). Autonomous attitude determination and control for planetary penetrator missions. *AIAA guidance, navigation, and control conference*. AIAA.

Ball, A. J. (2012). *Planetary landers, entry probes and penetrators*. Retrieved from Alpbach Summer School https://www.summerschoolalpbach.at/docs/2012/lectures/Ball_Alpbach_2012.pdf

Barraclough, S., Perkinson, M.-C., Peacocke, L., & Vijendran, S. (2015). Penetrator for the exploration of Europa. In *12th International planetary probe workshop*. Cologne, Germeny.

Biele, J., Ulamec, S., Hilchenbach, M., & Komle, N. (2010). In situ analysis of Europa ices by short-range melting probes. *Advances in Space Research*, 755–763.

Boyle, A. (2020, June 4). *Everything must boldly go! Defunct asteroid mining company's hardware put up for auction*. Retrieved from GeekWire https://www.geekwire.com/2020/everything-must-boldly-go-planetary-resources-hardware-auction-heats/

Brandner, A., & Koch, L. (2023). Speed-up measures and associated problems for large-scale DEM-models. *Berg Huettenmaenn Monatsh*, 287–293.
Bushnell, D. M. (2021). *Futures of deep space exploration, commercialization, and colonization: The frontiers of the responsibly imaginable.* Langley Research Center, NASA.
Cassady, J. (2014, August 4). *The ExoLance project and the search for life on Mars.* Retrieved from The Space Review https://www.thespacereview.com/article/2571/1
Chao, N., Cordes, J. A., Carlucci, D., DeAngelis, M., Marhevka, S., Lee, J., et al. (2010). The use of potting materials for electronic-packaging survivability in smart munitions. In *2010 International mechanical engineering conference and exposition.* Vancouver, British Columbia.
Chen, Y., Hu, S., Li, J.-H., Li, Q.-L., Li, X., Li, Y., et al. (2023). Chang'e-5 lunar samples shed new light on the Moon. *The Innovation Geoscience.*
Chung, J., Krauthammer, T., Zhang, Y., Mishra, N., Taghavi, A., Stone, M., Davidson, M. T., Taylor, A., Faraone, M., Patil, A., & Ilay, K. (2017). *Physics based prediction of unexploded ordnance penetration in granular materials.* SERDP.
Collinson, G. (2008). Planetary penetrators—The vanguard for the future exploration of the solar system. *Journal of the British Interplanetary Society,* 198–202.
David, L. (2015, September 18). *Asteroid-mining plan would bake water out of bagged-up space rocks.* Retrieved from Space.com https://www.space.com/30582-asteroid-mining-water-propulsion.html
Doyle, T. P. (2024, July 17). *NASA ends VIPER project, continues moon exploration.* Retrieved from NASA https://www.nasa.gov/news-release/nasa-ends-viper-project-continues-moon-exploration/
Driessen, P. P., & Henckens, T. (2016). Geological scarcity, market price trends, and future generations. *Resources Policy,* 102–111.
Emergen Research. (2022). *Space mining market size, share, trends, by type of asteroid (Type C, Type S, Type M, others), by phase (spacecraft design, launch, operation), by type of commodity (water and others), by application, and by region forecast to 2030.* Emergen Research.
ESA. (2025). *Report on the space economy 2025.* ESA.
Gao, Y., Ellery, A., Jaddou, M., Vincent, J., & Eckersley, S. (2007). Planetary micro penetrator concept study with biomimetric drill and sampler design. *IEEE Transactions on Aerospace and Electronic Systems.*
Gellert, R., & Clark, B. (2015). In situ compositional measurements of rocks and soils with the alpha particle X-ray spectrometer on NASA's mars rovers. *Elements,* 39–44.
Government of Japan. (2021). *Act on the promotion of business activities for the exploration and development of space resources (Act No. 83).*
Gowen, R., Smith, A., Winter, B., Theobald, C., Rees, K., Ball, A., et al. (2008). An update on MoonLITE. In *59th international astronautical congress.* International Astronautical Federation.
Grand Duchy of Luxembourg. (2017). *Loi sur l'exploration et l'utilisation des ressources de l'espace [Law on the exploration and use of space resources].*
Hecht, M., Hoffman, J., Rapp, D., J., M., SooHoo, J., Schaefer, R., et al. (2021). Mars oxygen ISRU experiment (MOXIE). *Space Science Reviews.*
Hepp, A., Palaszewski, B., Landis, G., Jaworske, D., Colozza, A., et al. (2014). In-situ resource utilization for space exploration: Resource processing, mission-enabling technologies,and lessons for sustainability on earth and beyond. *Propulsion and energy forum 2014.* American Institute of Aeronautics and Astronautics.
Hinze, W., vonFrese, R., & Saad, A. (2013). *Gravity and magnetic exploration, principles, practices, and applications.* Cambridge University Press.
Hopf, T., Kumar, S., Karl, W., & Pike, W. (2010). Shock protection of penetrator-based instrumentation via a sublimation approach. *Advances in Space Research,* 460–467.
Hughes. (1974). *Pioneer Mars surface penetrator mission.* NASA.
ISO. (2023). *Space systems—Space debris mitigation requirements.*
ispace. (2024). *ispace—About Us.* Retrieved from ispace https://ispace-inc.com/aboutus

JAXA. (2018, July 29). *Hayabusa2 information fact sheet.* Retrieved from Hayabusa2 https://www.hayabusa2.jaxa.jp/en/enjoy/material/factsheet/FactSheet_en_v2.31s.pdf

Kminek, G., Conley, C., Hipkin, V., & Yano, H. (2017). *COSPAR's planetary protection policy.* COSPAR.

Landers, V., Pathirana, O., Kornreich, E. B., & Aluvihare, H. (2024). Economical lunar sample return mission with soil penetration darts. In *75th International astronautical congress (IAC)* (pp. A3,IP,118,x89101). Intearnational Astronautical Federation.

Lauretta, D., Balram-Knutson, S., Beshore, E., Boynton, W., Aubigny, C., et al. (2017). OSIRIS-REx: Sample return from asteroid (101955) Bennu. *Space Science Reviews.*

Lawrence, D., Feldman, W., Goldsten, J., Maurice, S., Peplowski, P., et al. (2012). Evidence for water ice near Mercury's north pole from MESSENGER neutron spectrometer measurements. *Science,* 292–296.

Lorenz, R. D. (2011). Planetary penetrators: Their origins, history and future. *Advances in Space Research,* 403–431.

Luo, H., Li, Y., Fan, C., Wu, X., & Liu, G. (2021). Design and experimental research on buffer protection of high-g penetrator for deep space exploration. *Acta Astronautica,* 63–78.

Luxembourg Space Agency. (2025, March 07). *European Space Resources Innovation Center (ESRIC).* Retrieved from Luxembourg Space Agency https://space-agency.public.lu/en/expertise/space-directory/esric.html

McCarthy, N. (2021, February 19). *Exploring the red planet is a costly undertaking.* Retrieved from Statista https://www.statista.com/chart/24232/life-cycle-costs-of-mars-missions/

McCarty, J. L., Beswick, A. G., & Brooks, G. W. (1964). *Application of penetrometers to the study of physical properties of lunar and planetary surfaces.* NASA.

Mizutani, H., Fujimura, A., Tanaka, S., Shiraishi, H., & Nakjima, T. (2005). Lunar-A mission: Outline and current status. *Journal of Earth System Science,* 763–768.

Moulder, W. (2021, November 18). *Deployable antennas for small satellites.* Retrieved from https://ilp.mit.edu/sites/default/files/2021-11/Moulder-1100-deployableAntennasForSmallSats.pdf

Mueller, R., & Susante, P. J. (2012). A review of extra-terrestrial mining robot concepts. *Earth and Space 2012.* American Society of Civil Engineers.

Nakajima, T., Hinada, M., Mizutani, H., Saitoh, H., Kawaguchi, J., & Fujimara, A. (1996). Lunar penetrator program: LUNAR-A. *Acta Astronautica,* 111–119.

NASA. (2020). *The artemis accords.* Retrieved from https://www.nasa.gov/wp-content/uploads/2022/11/Artemis-Accords-signed-13Oct2020.pdf

NASA. (2025, April 07). *Commercial lunar payload services.* Retrieved from NASA https://www.nasa.gov/commercial-lunar-payload-services/

Noorda, A. Y., Polonskaya, K., & Burger, M. E. (2024). *Reimagining sustainable space mining with due diligence.* World Trade Institute.

Oleg, D., Schuster, G. L., Feng, X., Yongxiang, H., Hartmut, B., Jochen, L., & Zhengqiang, L. (2021). Grand challenges in satellite remote sensing. *Frontiers in Remote Sensing.*

Qian, S.-E. (2023). Overview of hyperspectral imaging remote sensing from satellites. *Advances in hyperspectral image processing techniques* (pp. 41–66).

Raj, C. G., Cato, M., Speller, N. C., Duca, Z., Putman, P., Epperson, J., et al. (2022). Icy moon penetrator organic analyzer post-impact component analysis. *Frontiers in Astronomy and Space Sciences.*

Raus, R., & Watt, Y. G. (2012). Analysis of state-of-the-art single-thruster attitude control techniques for spinning penetrator. *Acta Astronautica* (pp. 60–78).

Robertson, E. A., & Carson, J. M. (2015). ALHAT: Autonomous landing and hazard avoidance technology. *JSC technology poster session.* NASA Johnson Space Center.

Sanders, G. B. (2018). Space resources and mining: Current objectives, plans, and missions. *CASI ASTRO 2018.* Johnson Space Center.

Sanders, G. B., & Leinhenz, J. E. (2024). Update on NASA's ISRU development and mission plans for the artemis program. In *75th International astronautical congress (IAC).* International Astronautical Federation.

Sheridan, S., Barber, S., Morse, A., & Snodgrass, C. (2018). Penetrators as a deployment tool for Mass Spectrometer instrumentation. *European planetary science congress.* Open Research Online.

Smith, A., Crawford, I. A., Gowen, R. A., Ball, A. J., Barber, S. J., Church, P., et al. (2009). LunarEX—A proposal to cosmic vision. *Experimental Astronomy,* 711–740.

Smith, A., Crawford, I., Tong, V., Gowen, R., Coates, A., Griffiths, A., et al. (2010). *Lunar Net.* ESA.

Smrekar, S., Catling, D., Lorenz, R., Magalhaes, J., Moersch, J., Morgan, P., et al. (1999). Deep Space 2: The Mars microprobe mission. *Journal of Geophysical Research.*

Space Research Institute. (2002). *Robotic spacecraft mission to Mars. Brief description.* Retrieved from Mars 96 http://www.iki.rssi.ru/mars96/01_mars_e.htm

U.S. Government. (2019). *Orbital Debris Mitigation Standard Practices, November 2019 Update.*

U.S.Congress. (2015). *Commercial Space Launch Competitiveness Act—Pub. L. No. 114-90.*

UAE Space Agency. (2023, 03 31). *Policies and regulations.* Retrieved from UAE Space Agency https://space.gov.ae/en/policy-and-regulations

Ulaby, T., Long, D., Blackwell, W., Elachi, C., Fung, A., Ruf, C., et al. (2014). *Microwave radar and radiometric remote sensing.* University og Michigan Press.

United Nations. (1967). *Treaty on principles governing the activities of states in the exploration and use of outer space.*

United Nations. (1972). *Convention on international liability for damage caused by space objects.*

United Nations. (1975). *Convention on registration of objects launched into outer space.*

United Nations. (1979). *Agreement governing the activities of states on the moon and other celestial bodies.*

United Nations. (1994). *Agreement relating to the implementation of Part XI of the United Nations Convention on the Law of the Sea of 10 December 1982.*

US Army. (1946). *Comparative test of the effectiveness of large bombs against reinforced concrete structures.* The Air Proving Ground Command.

Völk, S., Ulamec, S., Biele, J., Hecht, M., Lell, P., Fleischmann, J., et al. (2018). Development and testing of a pyro-driven launcher for harpoon-based comet sample acquisition. *Acta Astronautica,* 218–228.

Wang, X., Qin, W., Bai, Y., & Cui, N. (2016). Precision navigation system design for Moon Penetrator. *Aircraft Engineering and Aerospace Technology,* 791–798.

WEF. (2024). *The $1.8 trillion opportunity for global economic growth.* World Economic Forum.

Werkheiser, N. (2014). Overview of NASA initiatives in 3D printing and additive manufacturing. *2014 DoD maintenance symposium.* NASA.

Winglee, R., Truitt, C., & Hoyt, R. (2012). *Sample return systems for extreme environments.* NASA.

Yu, Z. (2009). *Simulation of high-velocity penetration for rigid projectile into plain concrete target using discrete element method.* Virginia Polytechnic Institute and State University.

Zhou, J., Feng, J., Liu, J., Xu, C., Tian, Y., & Lu, D. (2024). Review on research and development of planetary penetrator. *Journal of Field Robotics.*

Sustainability in Space: Innovation, and International Cooperation

Saeed Nayeri, Al-Amin Dabo, F. Frank Alparslan, and Amin Hosseinian-Far

Abstract This chapter highlights the growing need for international collaboration and innovation to address the rising challenges of space sustainability. The chapter provides a sound overview of key technical, legal, economic, and social barriers, such as space debris, orbital congestion, regulatory gaps, and economic inequalities. Ethical concerns such as equitable access to space, responsible resource utilisation, and the environmental impact of space activities are also critically examined. The chapter argues that strengthening global partnerships, harmonising regulatory frameworks, and responsibly embracing technological advancements are essential to ensure a sustainable and inclusive future in space. The chapter also emphasises that only through cooperative, multi-stakeholder efforts can long-term space sustainability be effectively achieved.

Keywords Space · Sustainability · Innovation · Governance · International collaboration

S. Nayeri · A.-A. Dabo · F. F. Alparslan
Faculty of Business & Law, University of Northampton, Northampton, UK
e-mail: saeed.nayeri@northampton.ac.uk

A.-A. Dabo
e-mail: al-amin.dabo@northampton.ac.uk

F. F. Alparslan
e-mail: frank.alparslan@northampton.ac.uk

A. Hosseinian-Far (✉)
Sustainability Research Group, Hertfordshire Business School, University of Hertfordshire, Hatfield, UK
e-mail: a.hosseinian-far@herts.ac.uk

H. Jahankhani et al. (eds.), *Space Mining*, Space Law and Policy,
https://doi.org/10.1007/978-981-95-5543-7_9

1 Introduction

Space has always been viewed by mankind as a way to escape to new worlds and dimensions. Mythologies and historical events also indicate the desire of humans to explore beyond the boundaries of the Earth. Up until the last century, this was still seen as a dream rather than a tangible or practical reality. However, events such as the man on the moon in the 60s, the space race during the Cold War, and the very recent commercial revenue models associated with space exploration are making this dream a reality.

Conversely, life on Earth has proven to cause various dilemmas with respect to innovation adoption and cross-country collaborations within the boundaries of legal frameworks, and this offers a sound case study for future space-related activities. This is particularly imperative when considering the impacts of human-induced activities on Earth as our current primary habitat and operational landscape. One question that arises is what lessons could be learned from living on Earth, the impediments and challenges, and the innovations and technology adoptions for space as a new area of human activity.

This chapter focuses on these key elements and attempts to provide a review of some of the ethical dilemmas and challenges with respect to space sustainability, in particular, providing an overview of technological bottlenecks, regulatory constraints within the Earth's frameworks of legal systems, and pertinent socioeconomic challenges. The chapter provides an overview of how responsible innovation adoption and international cooperation could play a role in supporting our space ambitions. The chapter finally attempts to offer a number of reasonable recommendations for various stakeholders in this area for future space exploration activities.

2 Challenges and Ethical Dilemmas in Space Sustainability

The rapid expansion of space activities in recent years has brought forth unprecedented opportunities and challenges for space sustainability. As humanity ventures further into the cosmos, addressing these challenges becomes imperative to ensure that space remains a viable and equitable resource for future generations. The sustainability of space activities is not just a technical concern. It entails a complex web of legal, economic, and social factors. Technical advancements such as large satellite constellations, debris mitigation technologies, and in-situ resource utilisation provide some of the needed innovative solutions; however, they also introduce new risks and ethical dilemmas. Similarly, the lack of comprehensive legal frameworks, the high economic costs of sustainable practices, and social issues such as transparency, equity, and cultural preservation demand a multifaceted and inclusive approach in countries' efforts related to space exploration. The following sub-section provides a detailed examination of these challenges.

2.1 Technical Challenges

The emergence of new space technologies, such as large satellite constellations and in-orbit servicing capabilities, has created both opportunities and challenges with respect to the notion of sustainability. Although these innovations may offer solutions to some of the existing problems, they introduce new complexities in terms of increasing congestion in Earth's orbits. The exponential growth in satellite launches, particularly in low Earth orbit (LEO), has led to an unprecedented level of space traffic (Wilson & Vasile, 2023). This congestion increases the risk of collisions between active satellites and existing space debris, which can then potentially trigger a cascade of collisions; a phenomenon known as the Kessler Syndrome (Blechová et al., 2024). The deployment of mega-constellations (consisting of thousands of satellites aimed at providing global internet coverage), such as SpaceX's Starlink, has exacerbated this issue. For instance, in 2019, a near-collision occurred between a Starlink satellite and the ESA's Aeolus satellite, which further highlights the challenges of managing space traffic and the limitations of current coordination frameworks (Jee, 2019). Therefore, the threat from space debris and increased space activity is a serious issue that, if left unaddressed, could significantly jeopardise global markets and the commercial space economy.

Another technical challenge is the development and implementation of effective space debris mitigation and removal technologies. Efforts to mitigate space debris have included both passive and active removal technologies. Sace debris poses a significant technical challenge for space operations. With over 27,000 tracked objects currently in orbit, and rising annually, mitigating debris is increasingly difficult. The ethical issue arises when considering responsibility for cleanup efforts and the 'polluter pays' principle. According to this, nations and private entities that contribute to space debris must ethically bear the costs of mitigation to prevent harm to future missions and other space users (Carbajales-Dale & Murphy, 2023). The complexity of capturing and removing debris in the harsh space environment, coupled with the high costs associated with such missions, presents ongoing challenges for the space community.

Advancements in space exploration technologies allow missions to extract and utilise extraterrestrial resources. However, these technologies must balance innovation with planetary protection to prevent contamination of celestial bodies. The NASA Perseverance Rover, which landed on Mars in 2021, exemplifies the application of planetary protection protocols. The mission adheres to strict guidelines to avoid contaminating Martian environments, to ensure that scientific discoveries about potential life remain uncontaminated (Mason, 2021). Ethically, there is a duty to preserve the integrity of other planets, as highlighted in frameworks like the Outer Space Treaty (1967) (UNOOSA, 1979). Technical solutions must ensure that resource extraction does not compromise the potential for scientific discovery or harm native environment.

Lastly, technologies for human spaceflight, such as life-support systems, genetic enhancements, cognitive augmentation, AI integration, and automation in space mining improve efficiency, however, raise ethical concerns about accountability and decision-making, and also introduce ethical questions about human identity and fairness. Issues such as algorithmic bias, unintended consequences, and worker displacement must be addressed to ensure fairness and transparency in space-related activities (Yadav & Somu, 2024). Besides, unequal access to these enhancements can create disparities between 'enhanced' astronauts and those relying on traditional methods. For instance, discussions around CRISPR gene-editing technology for enhancing astronaut resilience against radiation exposure highlight potential disparities between 'enhanced' and traditional astronauts (Szocik et al., 2020).

2.2 Legal and Regulatory Challenges

One of the central challenges in space sustainability is striking the right balance between innovation adoption and ensuring responsible practices. The rapid pace of technological advancement in the space sector has led to new capabilities and opportunities. Nevertheless, this has also outpaced regulatory frameworks. Finding ways to encourage innovation and implementing effective sustainability measures requires careful consideration and adaptive policy-making (Lescrauwaet et al., 2022).

The common legal frameworks, such as the Outer Space Treaty (1967) (UNOOSA, 1967), Rescue Agreement (1968) (UNOOSA, 1968), The Liability Convention (1972) (UNOOSA, 1968), etc. are outdated in addressing modern related activities and respective challenges such as commercial space mining, space tourism, and mega-constellations. Therefore, they are increasingly strained by the rapid evolution of space technology and the emergence of new actors, including private companies. There is a growing need for updated and more comprehensive regulations to address issues such as space traffic management, debris mitigation, and the use of space resources (Pelton, 2019). Additionally, the voluntary nature of many existing guidelines, such as those for space debris mitigation, raises questions about their effectiveness in ensuring compliance across all space actors. Another key obstacle in this area is achieving international consensus on new binding agreements.

The principle of space as the "province of all humankind" conflicts with private ownership and national interests. Legal ambiguity over who has rights to space resources contributes to ethical dilemmas surrounding monopolisation and the exploitation of resources by wealthier nations and corporations (Khatwani, 2019). The diverse interests of spacefaring nations and the strategic importance of space assets often lead to protracted negotiations and difficulties in reaching agreements that are acceptable to all parties. Additionally, regulating fair competition in space activities is essential to prevent monopolies and ensure innovation benefits everyone. Ethical challenges related to 'space antitrust' regulations aim to balance corporate interests with public welfare and sustainability (Lucas-Rhimbassen, 2022). Moreover, the dual-use nature of space technology, where civilian technologies can be repurposed for military use, presents additional ethical challenges. Koskina (2023) emphasises the need for transparency in space operations to build trust among nations and prevent the misuse of space technologies. For instance, militarisation of space and the development of anti-satellite weapons raise significant ethical concerns. The 2019 Indian ASAT test demonstrated how such actions could compromise sustainability efforts by increasing debris in Low Earth Orbit (LEO) (Boley & Byers, 2021). Furthermore, the establishment of the U.S. Space Force in 2019 has intensified the militarisation of space. The potential for space to become a battlefield threatens the peaceful use of outer space and also poses risks to critical space infrastructure that supports global communication, navigation, and Earth observation systems. Yan (2023) highlights how this development challenges the notion of space as a peaceful domain and complicates international efforts to prevent an arms race in space. Balancing national security concerns with the commitment to peaceful exploration as well as prohibition or strict regulation of ASAT test although deemed challenging, are necessities for an appropriate space governance.

2.3 *Economic Challenges*

The economics of space sustainability also entails challenges of its own, primarily due to the high costs associated with adopting sustainable practices. These costs include designing satellites for end-of-life disposal, implementing green launch technologies, and conducting active debris removal missions. For instance, the European Space Agency (ESA)'s ClearSpace-1 mission, set to launch in 2025, aims to remove a single piece of space debris at an estimated cost of €120 million (Devlin, 2019). This mission represents a significant financial burden for a single debris removal operation, which indicates why commercial space actors might be reluctant to engage in similar efforts without financial incentives or mandatory regulations. Moreover, the high cost of developing sustainable technologies such as green propellants and reusable launch systems can deter smaller companies from adopting these practices. Although larger entities like SpaceX have successfully developed reusable rockets such as the Falcon 9, the upfront investment required for such technologies is substantial (London Economics, 2024). The economic feasibility of these sustainable practices becomes even more challenging for emerging space nations or smaller private companies without substantial financial backing.

Another key economic obstacle is the absence of clear models for valuing space infrastructure and quantifying the costs associated with space waste management. Unlike terrestrial industries, where waste management and environmental impact have established economic frameworks, space activities lack standardised methodologies for assessing the long-term benefits of sustainability. For instance, the economic impact of a single piece of space debris damaging an operational satellite can be catastrophic, which can then lead to other financial losses and also mission failures. However, assigning a clear monetary value to preventative measures such as debris mitigation is complex and often speculative (Jaehnichen, 2020). The difficulty in quantifying these costs complicates decision-making for both public agencies and private enterprises. Without robust economic models, space actors may prioritise short-term profitability over long-term sustainability, which then leads to continued debris accumulation and potential risks for future missions.

A significant ethical challenge arises from the tension between economic profitability and the long-term sustainability of space environments. Companies and nations often face the dilemma of pursuing immediate financial gains versus committing to sustainable practices that may not offer immediate returns. For instance, deploying large satellite constellations for global internet coverage, such as SpaceX's Starlink and Amazon's Project Kuiper, provides substantial economic benefits. However, these deployments contribute to orbital congestion and increase the risk of collisions and long-term debris issues. Balancing these economic interests with the ethical obligation to protect space for future generations remains a contentious issue (Bernat, 2020).

The development of lunar and Martian bases, such as those proposed under NASA's Artemis Program, requires significant economic investments. These missions aim to establish a sustainable human presence on the Moon, laying the groundwork for future Mars exploration. Such investments are offering scientific and economic advancement, however, they also raise ethical concerns about resource allocation. Critics argue that these funds could be directed towards pressing global issues on Earth, such as poverty alleviation, healthcare, and climate change mitigation and adaptation (Shagun & Anwer, 2024). For instance, the estimated cost of the Artemis Program exceeds $90 billion through 2025 (NASA, 2021), which prompts questions about whether such expenditures align with global equity and sustainable development goals. Ethical frameworks must weigh these investments against the broader impact on society, to ensure that space exploration does not exacerbate existing inequalities.

2.4 Social Challenges

Space missions funded by public resources should prioritise transparency and democratic accountability. Open communication and public engagement are essential to ensuring that space activities align with societal interests. For instance, NASA's Space Launch System (SLS) has faced significant public scrutiny due to cost overruns and delays, which have totalled over $20 billion since its inception (Howell, 2020). The controversy surrounding the SLS emphasises the importance of maintaining transparent communication with taxpayers who fund these missions. Public engagement can also encourage a sense of ownership and support for space programmes. Governments can promote democratic accountability by involving citizens in decision-making processes and providing clear information on mission objectives, costs, and risks. This ensures that space exploration reflects public priorities and ethical standards. However, transparency and accountability alone are not sufficient if the benefits of space exploration are not equitably distributed. The principle of "global equity" mandates that the benefits of space exploration be shared among all nations, not just those with advanced space programmes. However, the dominance of wealthier countries in space activities often exacerbates global inequalities (Rabitz, 2023). For instance, SpaceX's Starlink constellation provides global internet services; however, concerns remain about whether developing nations can afford access to these services. Additionally, the monopolisation of low Earth orbit (LEO) by large constellations limits opportunities for smaller nations to launch their own satellites.

To address these disparities, international cooperation frameworks such as the United Nations Office for Outer Space Affairs (UNOOSA) advocate for equitable access to space-derived technologies. However, achieving true equity requires tangible actions, such as capacity-building programmes and technology transfer to developing nations (Jakhu & Pelton, 2017). In addition, space exploration must also respect and preserve cultural heritage as humanity expands beyond Earth. The potential colonisation of other planets introduces ethical questions about preserving cultural heritage and respecting indigenous knowledge. The Moon, for instance, holds cultural significance for various societies, including indigenous communities. Future lunar missions must consider these cultural values to avoid exploitation and ensure respectful exploration practices (Balistreri & Umbrello, 2023). Furthermore, as exploration extends to Mars and beyond, ethical frameworks must address planetary protection protocols to avoid contaminating extraterrestrial environments. For instance, Mars missions must adhere to these protocols to prevent contamination that could compromise the discovery of potential indigenous Martian life (Conley & Rummel, 2010).

Beyond environmental and cultural preservation, the ethical implications of space colonisation also include issues of inequality and human enhancement. Space colonisation raises social challenges related to inequality and exploitation. If human enhancements, such as genetic modifications and cognitive augmentations, become standard for space colonists, it could create disparities between "enhanced" and "non-enhanced" populations (Szocik et al., 2020). Ethical frameworks must ensure that these advancements are accessible to all and do not exacerbate existing social inequalities.

Overall, the challenges and ethical considerations for space sustainability are complex and interconnected, and space across technical, legal, economic, and social domains. As space activities continue to grow, overcoming these challenges requires a comprehensive and collaborative approach. Technological advancements must be guided by ethical principles and supported by robust legal frameworks. Economic investments in space must balance profitability with long-term sustainability and global equity. And, social considerations such as transparency, democratic accountability, and cultural preservation must remain at the forefront of space exploration policies.

3 Harnessing Innovation and Global Collaboration

Addressing the challenges of space sustainability necessitates robust international cooperation and coordination. In fact, the global nature of space activities means that no single nation or entity can address these challenges alone. The need for shared responsibility, innovation, and regulatory frameworks stresses the importance of international communities working together. This section focuses on collaborative use cases, essential requirements, frameworks, facilitators, and challenges of international collaboration in space sustainability. The sub-sections below therefore examine the critical elements that underpin effective partnerships and highlight exemplary initiatives that demonstrate the power of collective action.

3.1 Key Use Cases

Examining key use cases of international collaboration provides valuable insights into how joint efforts can effectively address the challenges of space sustainability. The International Space Station (ISS) is a fine example, where NASA, ESA, Roscosmos, JAXA, and other agencies pool their resources and expertise to advance scientific research and technology development (Dominoni, 2024). The ISS facilitates continuous research in microgravity, and provides a platform for testing renewable energy technologies and efficient life-support systems. This model of sustained cooperation allows participating nations to share both the costs and benefits of space exploration and promote peaceful international relationships. Similarly, NASA and ESA collaborated on the Green Propellant Infusion Mission (GPIM), which successfully demonstrated the use of a hydroxylammonium nitrate-based fuel/oxidiser blend, known as AF-M315E, as a less toxic and more efficient alternative to traditional hydrazine propellants. Launched in June 2019 aboard a SpaceX Falcon Heavy rocket, GPIM tested this 'green' propellant in space, which showcased its potential to enhance spacecraft performance and to improve safety for launch personnel and reduce environmental hazards (Brabaw, 2019).

NASA's Open Innovation and Open-Source Platforms as another instance, leverage the principles of open collaboration to address complex space-related challenges through the involvement of diverse participants, including businesses, researchers, and the general public. These initiatives include the NASA Tournament Lab (NTL), which crowdsources solutions via partnerships with Harvard University and TopCoder; the NASA Space Apps Challenge, an annual hackathon inviting global problem-solving with NASA's open data; and NASA's GitHub Repository, which offers open-source space technology for external contributions. Noteworthy examples of this collaborative approach are the Astrobee Robot Project and OpenMDAO, an open-source framework for engineering optimisation. The programme enhances engagement, transparency, and efficiency and advances global participation and accelerates innovation cycles with real-world applications in space exploration and Earth sciences (Bailey, 2023). Utilising such open innovation platforms demonstrates how mutual goals and joint research can overcome geopolitical barriers. Together, these use cases demonstrate that pooling knowledge, resources, and technological capabilities leads to effective, sustainable outcomes in space exploration. These collaborations, for sure, advance technological innovation and also promote responsible and equitable use of space resources.

3.2 Requirements

International collaboration for encouraging sustainability in space is a complex task that necessitates several essential components to ensure efficiency and mitigate potential risks. The first critical element is establishing mutual goals and a shared vision, which forms the foundation for cohesive efforts and unified decision-making. As noted by Mariani et al. (2022), aligning the objectives of all entities involved helps to overcome cultural and operational differences, fostering trust and commitment. This unified vision ensures that collaborations are driven by shared sustainability interests, reducing misunderstandings and promoting long-term partnerships. Technological capabilities are another cornerstone of effective collaboration. According to Kaats and Opheij (2014), partners must possess the necessary infrastructure and research facilities to contribute meaningfully to joint projects. Equitable technological capabilities ensure balanced contributions and facilitate the exchange of advanced innovations, thus accelerating the development of sustainable solutions in space.

Legal and regulatory compliance is equally important in maintaining the legitimacy of collaborative efforts. As highlighted by Cinelli (2024), adherence to international treaties, such as the Outer Space Treaty, and national regulations is essential for avoiding legal conflicts and jurisdictional disputes. Clear legal frameworks provide guidelines that ensure ethical and lawful practices. In addition to legal compliance, data transparency and sharing are vital for promoting trust and subsequently enhancing collaboration. Dawes (2010) emphasises that robust mechanisms for data exchange help standardise information sharing and safeguard sensitive data while promoting transparency. Intellectual property (IP) management is another critical requirement for successful collaboration. Clear IP agreements, as noted by Bogers et al. (2013), help define ownership and usage rights, which then in turn curbs disputes and encourages innovation. Establishing guidelines for IP ensures that all parties are fairly recognised for their contributions and that potential conflicts are resolved efficiently. Collectively, these elements—mutual goals, technological readiness, legal compliance, data transparency, and IP management—provide a sound basis for achieving sustainable progress in space exploration and utilisation.

3.3 Collaboration Frameworks

International collaboration frameworks are essential for embracing innovation and achieving sustainability in space. Cooperation and integration models, as described by Rendleman and Faulconer (2010), emphasise pooling resources and sharing risks. These models range from basic cooperation, where partners work independently yet coordinate on specific tasks, to full integration, which involves joint decision-making and extensive resource-sharing. The United Nations Committee on the Peaceful Uses of Outer Space (COPUOS) has been at the forefront of promoting frameworks for international cooperation in space sustainability. In 2019, COPUOS adopted the Guidelines for the Long-Term Sustainability of Outer Space Activities, which sets a framework for policy and regulatory measures, safety of space operations, international cooperation, and scientific research (Martinez, 2021). These guidelines represent a significant step towards a global consensus on sustainable space practices. Regional organisations have also been contributory in promoting cooperation. The European Space Agency (ESA) has taken a leading role with initiatives such as the Zero Debris Charter, which commits signatories to achieving debris-neutral space activities by 2030 (ESA, 2024). This charter is another example that demonstrates the potential for regional leadership to drive global action on space sustainability.

Such collaboration reduces financial and operational burdens while ensuring equitable contributions from all entities involved. The relational space framework, highlighted by Bradbury-Huang et al. (2010), focuses on building trust and creating supportive environments. Trust is critical in long-term space projects, where uncertainty and high stakes are common. By promoting transparency and mutual respect, this framework strengthens partnerships and promotes innovation. Together, these models ensure that collaborative efforts are both efficient and resilient, and offer a foundation for sustained progress in space sustainability initiatives.

In addition to these models, other frameworks offer structured approaches for managing complex collaborations. The Living Labs framework, as discussed by Veeckman et al. (2013), supports innovation through real-world experimentation, which allows stakeholders to develop and test sustainable solutions in practical settings. This iterative approach aligns partners' interests and ensures that solutions are adaptable. The boundary objects framework, explained by Franco-Torres et al. (2020), uses shared tools and symbols to bridge institutional differences; this approach enhances communication and collaboration between entities with diverse goals or expertise. Finally, the Triple Helix framework, noted by Glittová and Šipikal (2022), integrates academia, industry, and government to drive multi-stakeholder innovation. This approach combines research, practical capabilities, and policy oversight, and promotes comprehensive and sustainable solutions. Together, these frameworks provide diverse yet complementary strategies for enabling effective collaboration in space sustainability.

3.4 Facilitators

Facilitating collaborative innovation in space sustainability relies on key enablers that improve cooperation, communication, and resource-sharing. Digital collaboration tools, such as European Aerospace Cluster Partnership (EACP) or ESA Open Space Innovation Platform (OSIP), provide vital platforms for real-time communication and global engagement. Abbate et al. (2019) highlight how these tools offer interactive spaces for idea-sharing and problem-solving, which make it easier for stakeholders across different regions to coordinate efforts. Features such as shared databases and collaborative design environments also support the co-creation of solutions and ensure broad dissemination of knowledge. These platforms are particularly beneficial for space sustainability projects involving geographically dispersed teams and promote transparency and a culture of openness. Partners can accelerate innovation, address challenges collectively, and maintain efficient communication by leveraging digital tools, as the responsible adoption of digital technologies ultimately streamlines the development of sustainable practices.

Another key facilitator is public–private partnerships (PPPs), which harness the strengths of both government and private sectors to drive large-scale innovations. Villani et al. (2017) emphasise how PPPs combine public resources, regulatory frameworks, and infrastructure with the agility, funding, and technological advancements of private enterprises. In space sustainability, these partnerships are crucial for developing clean energy technologies, waste management systems, and space exploration solutions. PPPs distribute risks and rewards, which results in ambitious and high-cost projects being more feasible. Additionally, policy support plays a vital role in enabling collaboration by creating favourable regulatory environments. These policies promote data sharing, intellectual property protections, and joint research funding, encourage international partnerships (Besednjak Valič et al., 2023). They also ensure that collaborative efforts align with global standards and sustainability goals. Accordingly, such global standards offer stability and predictability for long-term investments. The Space Sustainability Rating (SSR), developed by the World Economic Forum in collaboration with other organisations, is a fine instance of such supportive initiatives. The SSR aims to provide a standardised assessment of space missions' sustainability, encouraging responsible behaviour through transparency and recognition (Weforum, 2024). Another important initiative is the Promoting Space Sustainability Project, led by the United Nations Office for Outer Space Affairs (UNOOSA) with support from the United Kingdom Space Agency. This project focuses on raising global awareness of space sustainability and providing capacity-building services for emerging spacefaring nations (Citaristi, 2022). Such initiatives are crucial for ensuring that the benefits of sustainable space practices are accessible to all nations, regardless of their current space capabilities.

3.5 Barriers and Challenges

International collaboration for sustainability also faces significant challenges due to political, economic, technical, and cultural factors. One of the primary challenges is technological transfer constraints, where national security concerns and intellectual property rights limit the sharing of critical space technologies. Because many space technologies are dual-use (serving both civilian and military purposes), nations are often hesitant to share them, as they fear security risks (Rendleman & Faulconer, 2010). In addition, export control policies, such as restrictions on the international sale or transfer of technology, create obstacles for seamless cooperation between public and private entities from different countries (Mineiro, 2012). Another significant barrier is political and diplomatic challenges, where geopolitical tensions and differing national interests impede sustainable collaboration. Many spacefaring nations prioritise their strategic and economic objectives, which often lead to national exceptionalism and protectionism, and complicate the development of cohesive international sustainability policies (Aganaba-Jeanty, 2016).

Further complicating matters are institutional and regulatory barriers, where inconsistencies in national and international regulations create hurdles for cooperation. Different legal approaches to space sustainability, particularly regarding space debris mitigation and resource utilisation, make achieving consensus difficult (Malysheva & Hurova, 2022). Additionally, in emerging fields like space resource extraction, regulatory gaps highlight the urgent need for globally harmonised policies. Financial limitations also pose a major hurdle; economic and funding challenges delay or prevent collaborative efforts, as space missions demand significant investment. Many nations lack the financial resources required for sustained participation in international ventures, and economic disparities can lead to unequal contributions, which then undermine trust and collaboration (van Niekerk, 2020). Cultural and communication barriers exacerbate these issues, as language differences and varying organisational practices can impede effective collaboration. Misunderstandings, cultural biases, and a lack of mutual trust can stifle cooperation, particularly in long-term projects requiring close interaction (Ehrenfreund et al., 2010). Moreover, the need for improved data sharing and transparency in space activities is another critical challenge. Although there is growing recognition of the importance of space situational awareness, concerns about national security and commercial confidentiality often limit the free flow of information that is necessary for several collaboration opportunities such as effective space traffic management and collision avoidance (Oltrogge & Alfano, 2019).

Other challenges include environmental and sustainability concerns linked to the rapid growth of the space sector. Issues such as space debris, atmospheric pollution, and resource depletion threaten long-term sustainability. Balancing technological advancement with environmental preservation becomes difficult when national priorities conflict with global sustainability goals (Yap & Truffer, 2021). Public–private partnership (PPP) barriers also hinder collaboration, as bureaucratic inefficiencies and differing objectives between governments and private enterprises can slow progress. Effective PPPs require regulatory frameworks that align incentives and promote sustainable outcomes (Pinz et al., 2021). Finally, resistance to change and institutional inertia within organisations can prevent the adoption of innovative practices. Without strong leadership and administrative support, research institutions and agencies may struggle to implement necessary changes (Ávila et al., 2017). Addressing these multifaceted challenges requires integrated frameworks, cooperative policies, and the establishment of mutual trust among international stakeholders and private entities.

Overall, international collaboration for sustainability in space is a complex effort that relies on a delicate balance of shared goals, robust frameworks, effective facilitators, and the successful mitigation of challenges. Establishing mutual objectives, ensuring technological readiness, and adhering to legal and regulatory frameworks provide the foundations for cohesive and productive partnerships. Collaboration models such as Living Labs and the Triple Helix approach offer structured case studies and pathways to achieving these goals. Facilitators, including digital platforms, public–private partnerships, and supportive policies, enhance communication and resource-sharing and can make international cooperation more efficient. Key use cases, such as the ISS and clean energy initiatives, highlight the tangible benefits and innovations that arise from global partnerships. However, barriers such as technological transfer constraints, political challenges, and cultural differences continue to impede progress. Overcoming these obstacles requires integrated policies, mutual trust, and adaptive frameworks that can navigate the complexities of international cooperation, specifically because the rapid pace of innovations and technological developments in the space sector often outstrips the development of regulatory frameworks.

4 Recommendations

To bridge the gap between space sustainability challenges and international innovations and collaboration efforts, a multi-dimensional approach integrating policy development, technological advancements, economic incentives, and governance frameworks is necessary. The following recommendations aim to enhance cooperation and address the challenges associated with sustainable space activities.

4.1 Strengthening International Legal and Regulatory Frameworks

4.1.1 Updating Space Law and Policies

Existing treaties such as the Outer Space Treaty (1967) and the Liability Convention (1972) must be revised or supplemented with new agreements that address contemporary issues such as space debris management, commercial space mining, and space traffic coordination. The establishment of binding international laws, rather than voluntary guidelines, will ensure compliance among all stakeholders. This is one of the pressing requirements given the surge in the number and significance of space exploration projects globally.

4.1.2 Harmonisation of National Regulations

Countries should align their national space policies with international standards to prevent regulatory fragmentation. Establishing common licensing procedures and sustainability requirements for satellite operators will create a more predictable and cooperative environment for space activities. This may be considered a challenge, considering military secrets, and the dual use of space exploration technologies for military purposes. However, reassuring the internal community of the distinction between the two might provide a way forward.

4.1.3 Enhancing Space Traffic Management (STM) Coordination

The creation of an internationally governed STM system would improve the coordination of satellite launches, debris tracking, and collision avoidance. Organisations such as the United Nations Office for Outer Space Affairs (UNOOSA) and the International Telecommunication Union (ITU) should play a central role in standardising STM protocols. Other initiatives and innovative approaches could also be considered to ensure that the adopted technologies do not lead to longer-term unintended consequences.

4.2 Technological Innovations for Space Sustainability

4.2.1 Developing Advanced Debris Mitigation Technologies

Governments and private companies must invest further in active debris removal (ADR) technologies, such as robotic collection systems and laser-based deorbiting techniques. Establishing international funding mechanisms to support debris removal missions will encourage innovation and participation from both developed and emerging space nations. Although this might not seem vital, it is in fact imperative to start such investments as early as possible to prevent any future technological bottlenecks for future advancements in space exploration.

4.2.2 Promoting Green Propulsion and Sustainable Manufacturing

The adoption of non-toxic, eco-friendly propellants, such as AF-M315E, and the use of recyclable materials in satellite construction should be incentivised. Research collaborations between academia, industry, and government can accelerate the development and implementation of sustainable technologies. The notion of reuse, where applicable, should also be embraced. More research will also be required on the impact of space launches on fauna and flora, particularly marine ecosystems, to identify mitigating solutions.

4.2.3 Ensuring Ethical AI Integration in Space Operations

AI-based decision-making in space mining, navigation, and mission planning must adhere to transparent, fair, and accountable ethical frameworks. Developing global AI governance policies will help mitigate issues related to algorithmic bias and workforce displacement. There also have to be considerations with respect to artificial general intelligence (AGI) for use in future manned or unmanned spacecraft.

4.3 Economic Incentives and Financial Mechanisms

4.3.1 Establishing Market-Based Incentives for Sustainability

Introducing carbon credit-like systems for space sustainability, where companies receive credits for sustainable practices such as debris removal, could promote responsible behaviour. Tax benefits and financial grants for sustainable space technologies should also be explored. The lessons learnt from practices on Earth can provide guidelines to curb greenwashing.

4.3.2 Encouraging Public–Private Partnerships (PPPs)

Governments should incentivise partnerships between space agencies, private enterprises, and research institutions to share the financial burden of sustainable space initiatives. PPPs can be instrumental in funding large-scale sustainability projects such as on-orbit servicing and in-space recycling programmes.

4.4 Enhancing International Collaboration and Trust

4.4.1 Expanding Global Space Sustainability Initiatives

Programmes such as the Space Sustainability Rating (SSR) and the Zero Debris Charter should be expanded to include a broader range of stakeholders. Capacity-building initiatives should be prioritised to ensure that low- and middle-income nations have access to technology, training, and funding for sustainable space activities.

4.4.2 Fostering Transparency and Data Sharing

A centralised, open-access space situational awareness (SSA) database should be developed to enable global cooperation in tracking space debris and preventing collisions. National security concerns should also be balanced with the collective need for transparency in space operations.

5 Conclusion

Space sustainability is a critical global challenge that demands a multidisciplinary and cooperative approach. This chapter has examined the technical, legal, economic, and social challenges that hinder the sustainable use of outer space. We highlighted some of the ethical dilemmas associated with technological advancements, regulatory gaps, economic constraints, and social inequities.

International cooperation remains the basis of addressing these challenges. The success of collaborative initiatives, such as the International Space Station (ISS) and the Space Sustainability Rating (SSR), demonstrates the potential of global partnerships in promoting responsible space activities. However, political tensions, regulatory disparities, and financial limitations continue to obstruct effective cooperation.

To achieve long-term sustainability in space, nations and stakeholders must commit to updating legal frameworks, to promote technological innovations, implement economic incentives, and reinforce international governance mechanisms. Strengthening transparency, accountability, and inclusivity in space activities will be essential in ensuring that outer space remains a shared and accessible resource for future generations.

References

Abbate, T., Codini, A. P., & Aquilani, B. (2019). Knowledge co-creation in open innovation digital platforms: Processes, tools and services. *Journal of Business & Industrial Marketing, 34*(7), 1434–1447. https://doi.org/10.1108/JBIM-09-2018-0276

Aganaba-Jeanty, T. (2016). Space sustainability and the freedom of outer space. *Astropolitics, 14*(1), 1–19. https://doi.org/10.1080/14777622.2016.1148463

Ávila, L. V., et al. (2017). Barriers to innovation and sustainability at universities around the world. *Journal of Cleaner Production, 164*, 1268–1278. https://doi.org/10.1016/j.jclepro.2017.07.025

Bailey, G. L. (2023). *Propelling NASA closer to the moon and mars with open innovation, NASA*. Retrieved January 2, 2025, from https://www.nasa.gov/general/propelling-nasa-closer-to-the-moon-and-mars-with-open-innovation/

Balistreri, M., & Umbrello, S. (2023). Modifying the environment or human nature? What is the right choice for space travel and mars colonisation? *NanoEthics, 17*(1), 5. https://doi.org/10.1007/s11569-023-00440-7

Bernat, P. (2020). Orbital satellite constellations and the growing threat of Kessler syndrome in the lower earth orbit. *Inżynieria Bezpieczeństwa Obiektów Antropogenicznych* [Preprint], (4). https://doi.org/10.37105/iboa.94

Besednjak Valič, T., et al. (2023). Key policy mechanisms supporting the University–Industry collaboration in the Danube region: Case study of academic HPC centres and SMEs. *European Journal of Management and Business Economics, 32*(5), 509–524. https://doi.org/10.1108/EJMBE-09-2022-0283

Blechová, A., Harašta, J., & Kasl, F. (2024). From space debris to space weaponry: A legal examination of space debris as a weapon. In *2024 16th International conference on cyber conflict: Over the horizon (CyCon)* (pp. 263–279). IEEE. https://doi.org/10.23919/CyCon62501.2024.10685616

Bogers, M., Bekkers, R., & Granstrand, O. (2013). Intellectual property and licensing strategies in open collaborative innovation (pp. 1204–1224). In *Digital rights management*. IGI Global. https://doi.org/10.4018/978-1-4666-2136-7.ch059

Boley, A. C., & Byers, M. (2021). Satellite mega-constellations create risks in low Earth orbit, the atmosphere and on Earth. *Scientific Reports, 11*(1), 10642. https://doi.org/10.1038/s41598-021-89909-7

Brabaw, K. (2019). *NASA's 'Green' fuel satellite passes its first test, Space.com*. Retrieved January 2, 2025, form https://www.space.com/green-propellant-satellite-success.html

Bradbury-Huang, H., et al. (2010). Relational space and learning experiments: The heart of sustainability collaborations. *Research in Organizational Change and Development, 18*, 109–148. https://doi.org/10.1108/S0897-3016(2010)0000018008

Carbajales-Dale, M., & Murphy, T. W. (2023). The environmental and moral implications of human space travel. *Science of The Total Environment, 856*, 159222. https://doi.org/10.1016/j.scitotenv.2022.159222

Cinelli, C. (2024). *Regulation of outer space*. Routledge. https://doi.org/10.4324/9781003512677

Citaristi, I. (2022). United Nations Office for Outer Space Affairs—UNOOSA. In *The Europa directory of international organizations 2022* (pp. 247–248). Routledge. https://doi.org/10.4324/9781003292548-53

Conley, C. A., & Rummel, J. D. (2010). Planetary protection for human exploration of Mars. *Acta Astronautica, 66*(5–6), 792–797. https://doi.org/10.1016/j.actaastro.2009.08.015

Dawes, S. S. (2010). Stewardship and usefulness: Policy principles for information-based transparency. *Government Information Quarterly, 27*(4), 377–383. https://doi.org/10.1016/j.giq.2010.07.001

Devlin, H. (2019). *European Space Agency to launch space debris collector in 2025, Guardian*. Retrieved January 2, 2025, from https://www.theguardian.com/science/2019/dec/09/european-space-agency-to-launch-clearspace-1-space-debris-collector-in-2025

Dominoni, A. (2024). Design for sustainability in space: Future nature. In *Design for sustainability and inclusion in space* (pp. 25–48). Springer. https://doi.org/10.1007/978-3-031-56004-0_2

Ehrenfreund, P., et al. (2010). Cross-cultural management supporting global space exploration. *Acta Astronautica, 66*(1–2), 245–256. https://doi.org/10.1016/j.actaastro.2009.05.030

ESA. (2024). *Twelve countries sign the Zero Debris Charter, ESA*. Retrieved January 2, 2025, from https://www.esa.int/Space_Safety/Space_Debris/Twelve_countries_sign_the_Zero_Debris_Charter#:~:text=The%20Zero%20Debris%20Charter%20is,Seville%20meeting%20in%20November%202023

Franco-Torres, M., Rogers, B. C., & Ugarelli, R. M. (2020). A framework to explain the role of boundary objects in sustainability transitions. *Environmental Innovation and Societal Transitions, 36*, 34–48. https://doi.org/10.1016/j.eist.2020.04.010

Glittová, K., & Šipikal, M. (2022). University science parks as an innovative tool for university-business cooperation. *European Conference on Innovation and Entrepreneurship, 17*(1), 648–656. https://doi.org/10.34190/ecie.17.1.399

Howell, E. (2020). *Over budget, behind schedule: NASA's SLS megarocket faces congressional review, Space.com*. Retrieved January 2, 2025, from https://www.space.com/nasa-sls-megarocket-cost-schedule-oig-report-2020.html

Jaehnichen, T. (2020). The dynamics of economic action and the problems of its social embedding—Ethical challenges in view of the nascent commercial use of outer space. *HTS Teologiese Studies/Theological Studies, 76*(1). https://doi.org/10.4102/hts.v76i1.5996

Jakhu, R. S., & Pelton, J. N. (2017). *Global space governance: An international study*. In R. S. Jakhu & J. N. Pelton (Eds.). Springer International Publishing. https://doi.org/10.1007/978-3-319-54364-2

Jee, C. (2019). *One of SpaceX's Starlink satellites almost collided with a weather satellite, MIT technology review*. Retrieved January 2, 2025, from One of SpaceX's Starlink satellites almost collided with a weather satellite

Kaats, E., & Opheij, W. (2014). *Creating conditions for promising collaboration*. Springer. https://doi.org/10.1007/978-3-642-41443-5

Khatwani, N. (2019). Common heritage of mankind for outer space. *Astropolitics, 17*(2), 89–103. https://doi.org/10.1080/14777622.2019.1638679

Koskina, A. (2023). The use of AI weapons in outer space: Regulatory challenges. *Artificial Intelligence and Normative Challenges, 2*, 237–254. https://doi.org/10.1007/978-3-031-41081-9_13

Lescrauwaet, L., et al. (2022). Adaptive legal frameworks and economic dynamics in emerging technologies: Navigating the intersection for responsible innovation. *Law and Economics, 16*(3), 202–220. https://doi.org/10.35335/laweco.v16i3.61

London Economics. (2024). *Crouching rivals, not-so-hidden dragon: SpaceX and the future of launch competition—Part 1, London economics*. Retrieved January 2, 2025, from https://londoneconomics.co.uk/blog/publication/crouching-rivals-not-so-hidden-dragon-spacex-and-the-future-of-launch-competition-part-1/

Lucas-Rhimbassen, M. (2022). The COST of joining legal forces on a celestial body of law and beyond: Anticipating future clashes between corpus Juris Spatialis, Lex Mercatoria, Antitrust and ethics. *Space Policy, 59*, 101445. https://doi.org/10.1016/j.spacepol.2021.101445

Malysheva, N. R., & Hurova, A. M. (2022). Long-term sustainability of space activities: New challenges for international and national space law. *Kosmìčna nauka ì tehnologìâ, 28*(6), 63–73. https://doi.org/10.15407/knit2022.06.063

Mariani, L., et al. (2022). Achieving sustainable development goals through collaborative innovation: Evidence from four European initiatives. *Journal of Business Ethics, 180*(4), 1075–1095. https://doi.org/10.1007/s10551-022-05193-z

Martinez, P. (2021). The UN COPUOS guidelines for the long-term sustainability of outer space activities. *Journal of Space Safety Engineering, 8*(1), 98–107. https://doi.org/10.1016/j.jsse.2021.02.003

Mason, C. (2021). *Could humans have contaminated Mars with life?, BBC*. Retrieved January 2, 2025, from https://www.bbc.co.uk/future/article/20210510-could-the-perseverance-rover-have-carried-life-to-mars

Mineiro, M. (2012). *Space technology export controls and international cooperation in outer space*. Springer. https://doi.org/10.1007/978-94-007-2567-6

NASA. (2021). *NASA's management of Artemis mission, NASA*. Retrieved January 2, 2025, from https://oig.nasa.gov/docs/IG-22-003.pdf

van Niekerk, A. (2020). Inclusive economic sustainability: SDGs and global inequality. *Sustainability, 12*(13), 5427. https://doi.org/10.3390/su12135427

Oltrogge, D. L., & Alfano, S. (2019). The technical challenges of better space situational awareness and space traffic management. *Journal of Space Safety Engineering, 6*(2), 72–79. https://doi.org/10.1016/j.jsse.2019.05.004

Pelton, J. N. (2019). A path forward to better space security: Finding new solutions to space debris, space situational awareness and space traffic management. *Journal of Space Safety Engineering, 6*(2), 92–100. https://doi.org/10.1016/j.jsse.2019.04.005

Pinz, A., Roudyani, N., & Thaler, J. (2021). Public-private partnerships as instruments to achieve sustainability-related objectives: The state of the art and a research agenda. In *Sustainable Public Management* (pp. 3–24). Routledge. https://doi.org/10.4324/9781003202479-2

Rabitz, F. (2023). Space resources and the politics of international regime formation. *International Journal of the Commons, 17*(1), 243–255. https://doi.org/10.5334/ijc.1274

Rendleman, J. D., & Faulconer, J. W. (2010). Improving international space cooperation: Considerations for the USA. *Space Policy, 26*(3), 143–151. https://doi.org/10.1016/j.spacepol.2010.06.008

Shagun, S., & Anwer, N. (2024). Space-based solar power: Legal frameworks and sustainable development perspectives. *Discover Applied Sciences, 6*(12), 637. https://doi.org/10.1007/s42452-024-06259-5

Szocik, K., Norman, Z., & Reiss, M. J. (2020). Ethical challenges in human space missions: A space refuge, scientific value, and human gene editing for space. *Science and Engineering Ethics, 26*(3), 1209–1227. https://doi.org/10.1007/s11948-019-00131-1

UNOOSA. (1967). *2222 (XXI). Treaty on principles governing the activities of states in the exploration and use of outer space, including the moon and other celestial bodies, outer space treaty (OST)*. Retrieved January 6, 2025, from https://www.unoosa.org/oosa/en/ourwork/spacelaw/treaties/outerspacetreaty.html

UNOOSA. (1968). *Agreement on the rescue of astronauts, the return of astronauts and the return of objects launched into outer space, rescue agreement*. Retrieved January 6, 2025, from https://www.unoosa.org/oosa/en/ourwork/spacelaw/treaties/introrescueagreement.html

UNOOSA. (1979). *Agreement governing the activities of states on the moon and other celestial bodies, moon agreement*. Retrieved January 6, 2025, from https://www.unoosa.org/oosa/en/ourwork/spacelaw/treaties/intromoon-agreement.html

Veeckman, C., et al. (2013). Linking living lab characteristics and their outcomes: towards a conceptual framework. *Technology Innovation Management Review, 3*(12), 6–15. https://doi.org/10.22215/timreview/748

Villani, E., Greco, L., & Phillips, N. (2017). Understanding value creation in public-private partnerships: A comparative case study. *Journal of Management Studies, 54*(6), 876–905. https://doi.org/10.1111/joms.12270

Weforum. (2024). *The world's first sustainability rating system for space exploration, Weforum*. Retrieved January 2, 2025, from https://www.weforum.org/impact/world-s-first-space-sustainability-rating-launched/

Wilson, A. R., & Vasile, M. (2023). The space sustainability paradox. *Journal of Cleaner Production, 423*, 138869. https://doi.org/10.1016/j.jclepro.2023.138869

Yadav, R. K., & Somu, S. C. (2024). Ethical and societal implications of artificial intelligence in space mining. In *2024 International conference on trends in quantum computing and emerging business technologies* (pp. 1–5). IEEE. https://doi.org/10.1109/TQCEBT59414.2024.10545295

Yan, Y. (2023). Anti-weaponization of outer space for maintaining long-term sustainability of outer space activities. *Space Policy, 63*, 101519. https://doi.org/10.1016/j.spacepol.2022.101519

Yap, X.-S., & Truffer, B. (2021). Opportunities and threats of the rapidly developing Space sector on sustainability transitions: Towards a research agenda. *GEIST-Geography of Innovation and Sustainability Transitions,* [Preprint] (02).

The manufacturer's authorised representative in the EU is Springer Nature Customer Service Centre GmbH, Europaplatz 3, 69115 Heidelberg, Germany. If you have any concerns regarding our products, please contact ProductSafety@springernature.com

Printed and bound by CPI Group (UK) Ltd, Croydon, CR0 4YY
21/09/2026
02219990-0001